The Metronomic Society

The Metronomic Society

NATURAL RHYTHMS AND HUMAN TIMETABLES

Michael Young

HARVARD UNIVERSITY PRESS
CAMBRIDGE, MASSACHUSETTS
1988

This book is printed on acid-free paper, and its binding materials
have been chosen for strength and durability.

Library of Congress Cataloging-in-Publication Data
Young, Michael Dunlop, 1915–
    The metronomic society.

    Includes index.
    1. Time—Social aspects. 2. Cycles. I. Title.
HM299.Y68   1988   302   87-30663
ISBN 0-674-57195-9 (alk. paper)

For Edith Dunlop Young

Contents

Preface

This book began in the form of three lectures which I gave in the Sociology Department of Harvard University, at William James Hall, in March 1985. I am grateful to Professor Daniel Bell for inviting me to give them and to William James himself for giving me so many ideas, which it is not his fault if I have misinterpreted. The audience was a general one, primarily but not exclusively made up of sociologists, and the book is also intended for such an audience. As in the lectures, my hope is that by using the three biological analogies I have chosen I may be able to gain a little distance from the mysterious subject of society, without making it seem smaller.

I have on many topics gone beyond my own knowledge and so have consulted many people more expert than I am. I am particularly grateful, among my fellow sociologists, for the advice I have had from Daniel Bell; Paul Barker (if he does not mind my calling him a sociologist); Ian Cullen of University College, London; Mary Douglas, now of Princeton University; Peter Marris of UCLA; Jack Goody of Cambridge University; Hilda and Leo Kuper of UCLA; Adam Kuper of Brunel University (and his excellent seminar); Murray Melbin of Boston University; Pierre van den Berghe of the University of Washington; and my colleagues Johnston Birchall, Tom Schuller, and Simon Szreter, of the Institute of Community Studies. Among biologists and physiologists I have had valuable help from Monica Allfrey of Taunton; John Maynard Smith of Sussex University; Patrick Bateson of Cambridge University; John Cloudsley-Thompson of Birkbeck College; Michael Marmot of University College, London; Colin Patterson of the Natural History Museum, London; Jim Waterhouse of Manchester University; E. O. Wilson of Harvard; and J. Z. Young of Oxford; among physicists, from David Layzer of Harvard and John Ziman of Imperial College; among philosophers, from Stanley Cavell of Harvard, Philippa Foot of UCLA, Geoffrey Lloyd of Cambridge University, Tony Long of Berkeley, and Ruth Padel of Birkbeck College; among psychologists, from Jeffrey Adams of The Open College, Christopher Badcock of the London School of Economics, Marie Jahoda of Sussex University, and Jerome Kagan of Harvard; and among geographers, from Peter Hall of Reading University and the University of California, and Tommy Carlstein of the University of Lund.

In addition, I am grateful to the people who attended a seminar at the Institute of Community Studies to discuss a draft of this book—Peter Willmott (the Institute's chairman), James Douglas, Ronald Dore, Geoffrey Hawthorn, Peter Laslett, Charles Madge, and Roger Warren-Evans. I am only sorry that I met Julius Fraser so late in my writing. He is *the* student of time, and this would have been a better book if I had met him earlier, along with that other outstanding student of the subject, G. J. Whitrow.

More personally, I am most especially grateful to my wife, Sasha, for her continuing help and patience throughout what at the moment seems to have been a long travail. My sons, David and Toby, have been most stimulating. I also have a considerable debt to friends for their comments and support—to Clark Abt, Vincent Brome, Simon Folkard, Stephen Graubard, Donald MacRae, Martha Nelson, Jean Pool, Marianne Rigge, Eirlys Roberts, Edward Shils, Martin Shuttleworth, Prudence Smith, Gwyneth Vernon, Vesna Vucinic, Maysie Webb, Patricia Williams, and Joan Young—as well as to Jim Witte of Harvard, who acted as my research assistant while I was there; Lesley Cullen, my personal assistant, who was invaluable in digging out sources and checking references; and Jennifer Snodgrass, my editor. I was also much helped by the American Academy of Arts and Sciences in

Cambridge, Massachusetts; John Friedmann, of the School of Architecture and Planning at UCLA; and Michael Teitz, of the Department of City and Regional Planning at the University of California, Berkeley, who between them gave me precious months away from my London office in which to write. I am also grateful to the Leverhulme Trustees for financial assistance.

I am, above all, once more in debt to Sue Chisholm, who has typed draft after draft with marvelous skill and patience, and to Wyn Tucker, who has kept the Institute going while the writing haltingly proceeded.

My mother, Edith Dunlop Young, first aroused my interest in the general subject. She used to recite to me at bedtime the passage from the *Bhagavad Gita* that serves as an epigraph for this work. She died while the book was in manuscript. I have dedicated the book to her.

The epigraphs to the chapters are from the greatest master of time, Shakespeare. Almost as needless to say, all the mistakes are mine.

Never the spirit was born
The spirit shall cease to be never
Never was time it was not
End and beginning are dreams
Birthless and deathless and changeless
Remaineth the spirit for ever
Death hath not touched it at all
Dead though the house of it seems

Bhagavad Gita

ONE

The Cyclical and the Linear

Like as the waves make towards the pebbled shore,
So do our minutes hasten to their end;
Each changing place with that which goes before,
In sequent toil all forwards do contend.

Sonnet 60

The experimental English school where I was once a pupil, at Dartington Hall in the reassuringly unchanging County of Devon, was sired by Rabindranath Tagore, the Indian poet, out of John Dewey, the American philosopher. Tagore recommended that all the children undergo what he called night education. We were supposed to take ourselves off into the dark woods and fields, first with an adult and then, as our primitive terror lessened, on our own. Along the dim pathways we should be able to concentrate better on smelling and touching and, above all, listening to the other creatures of the night; we would learn to brave the cold wind and the ghosts who creep about in the bushes and sway in the branches of the trees. With night as tutor (Tagore believed) we might learn to appreciate a different aspect, which is also a different time, of nature, and of our own natures.

His teaching has not left any marked impression on the educational systems of England or the world; nor on me, until I recently spent a summer night with Nurse Bellamy and twenty of her elderly patients in the Nightingale Ward of a general hospital in southern England. I was collaborating with Jeffrey Adams and Simon Folkard on a study of nurses on the nightshift.[1] Nurse Bellamy was supposed to pee into a bottle at regular intervals during the night so that a potassium count of her urine could be made for us in the hospital laboratory. I was there to observe the work she did between times:

Soon after 10 o'clock she wheeled her mobile desk out of the office into the middle of the ward. Lights out. Except as vague shapes, she could not see her patients but, with her trained ear, she could monitor them continuously. In the growing hush broken only by the occasional cough or moan, she was listening for a catch in anyone's breathing. At any falter in the rhythmic sounds coming from each bed, she jumped up noiselessly to investigate. In between checks she whispered in low tones to me but without any lapse in her concentration on her patients.

The transformation which gradually gave my eyes back their sight came soon after the last owl stopped hooting and the last foxes stopped barking in the hospital grounds. As the daytime birds outside struck up their dawn chorus, the high, uncurtained windows began to take shape. If light travels faster than anything else in the universe, it did not seem like it then. In this little bit of the universe it seemed to be traveling more slowly than anything else. It did not seem as though light intensity was doubling every three minutes, as it does.[2] As the hospital with me in it gradually slid over on its side toward the sun, the barely perceptible pink light at first touched only the high clouds seen through the top of the window. Then it crept downwards. It was like an invisible hand stroking each pane of glass into the likeness of an x-ray plate, before the shadows in it take shape as bones. I could not notice the change in the look of the panes while it was happening. I could notice, continuously, not what was happening but that it *had* happened by throwing my mind back to a moment before when the panes were darker. I evidently had a large, if temporary, mental store of panes of different shades which my mind's eye could compare with what my actual eyes showed me.

As each bed and the hump under each coverlet emerged from its night shroud and the eyelids and the veined hands became suffused with the faint glow of the new day, this was the signal for Nurse Bellamy to wheel the desk back into the office and start her first ward round. Looking as fresh as if she also had been asleep, as she moved from one bed to another she knew which of the patients had feared they might never wake, and which had lost most of their faculties and been forced back on the hearing that she told me is the last sense to go, the last stronghold of people near death. She knew who would be specially reassured by the familiar sequence of morning sounds—the rustle of sheets as people pulled their arms from under the bedclothes so that she could take their pulses, the muffled tap as she slipped the thermometer back into its container on the wall, the swish of her skirt as she moved on to the next bed.

In this setting the most humdrum of events—the beginning of a new day—was for once not humdrum at all but startling. In retrospect, I saw how complete the change had been. In a few minutes the humps had come to life as pale faces with rheumy eyes, propped up against the pillows. And how gradual it had been, a gliding gradualness. There were fixed points in space, like the x-ray windows, but no fixed points in time. If there were any present moments, they were moments less of being than of becoming. After the day had established itself, space in the ward could declare itself as all of a piece and all at once, in the manner of geometry, but in the transition from darkness the rising sun had momentarily converted the all-of-a-piece and the all-at-once of night into the bit-by-bit of dawn, more like the calculus than geometry. Morning arrived with the grace of a Tai-Chi dancer. Locke defined time as perishing;[3] but in this hospital where people were on the point of perishing it seemed more like a murmurous revival.

Observing all this, I remembered not only the window panes of a moment ago but the Tagore of a little longer ago. It seemed to me that Nurse Bellamy had benefited from a different kind of night education from his, but a very real one; whereas I had learned so little from his teaching that, while being rolled more than 20,000 times on my axis, I had braced and blindfolded myself so tightly that until that hospital night I had hardly appreciated a single dawn. But if I have insulated myself from the cycles of night and day, and even of summer and winter, have I been the only traveler who has taught himself so skillfully not to see?

Stressing the Cyclical

In this book I am going to unbrace myself a bit and try to become more aware of other dawns and other recurrences into which we are bound by the nature of our revolving environment, and into which we have bound ourselves by what we have done with it. For this purpose I need a word to describe the phenomenon, and "cyclical" seems to do well enough. In this chapter I want to stress just this particular element in behavior. Even if it goes against the grain of modern thinking to admit it, I want to draw attention to the crucial survival of the cyclical. The term is contrasted with "linear" (used in a nonmathematical sense). The cyclical keeps things the same by reproducing the past and the linear makes things different by introducing novelty. These are the two dimensions of time (as I call them) or the two sorts of change which I shall compare with each other.

I can illustrate what I mean by picking out the cyclical elements in Nurse Bellamy's work. She follows the same sequence of actions each morning. She wheels her desk back into the office (Action A) and then goes on her round—using that word, too—to the first bed, where she goes through a certain sequence (Action B), and then on to the second bed, where she repeats the sequence (Action C), and so forth. Action A is a repeat of what has happened on other mornings, perhaps on thousands of other mornings in thousands of other hospitals; Actions B and C likewise, whereas Action C is also a repeat of Action B on the same morning. I shall argue later on that the more often Nurse Bellamy, or anyone else, follows the same sequence of cyclical changes, the more entrenched the sequence becomes. Cycles build up their own momentum. The motion is carried over from one occasion to another. Cycles are the dynamic feature of the social structure which give it stability.

I am stressing the cyclical because of the common tendency to do the opposite. The linear—the fact that Bellamy's C is not the same as B or that she never performs any sequence (however habitual) in exactly the same way—can be noticed to the exclusion of the cyclical. The attitude which fits modern society is less that everything you see you are seeing again than that everything you see you are seeing for the first and last time. It has become more evident that the river flows into the ocean than that in the ocean, though the waves undulate, the body of water does not move. When asked how he could repeat the same brush design on his pots for thirty years, Shoji Hamada, the great Japanese potter, was in the modern idiom when he replied that his designs were never quite the same. "Even if I tried to repeat, the pigment, brush, my arm and the thought would be different."[4] When the linear so much overshadows the cyclical, then by hiding the vital, iterative part of its own constitution the social structure becomes that much more difficult to comprehend.

The modern view seems to be that the cyclical may have been perfectly acceptable in some ancient societies in which the wheel and the mandala were appropriate symbols for time, but not for us. We can read, but only as if from a great distance, that in India (according to Mircea Eliade) an elaborate system of longer cycles was added to the natural recurrences of the days, months, and years, with 360 ordinary years making a divine year, and 12,000 divine years forming another repeating cycle.[5] It is difficult for us to credit such an ardent belief that everything repeated itself: it is foreign to our modern view of a linear history in which nothing repeats itself. It is also perplexing that in many societies which postulated an extratemporal moment of creation the original rituals had to be continually re-created and the old myths continually retold in order to restore the original life forms. "Every creation repeats the pre-eminent cosmogonic act, the Creation of the world."[6] In Christianity the eternal return lives on in the idea that Easter is a time for total renewal as well as for Christ's resurrection.[7]

Such notions can be brushed aside if they refer only to the past, and this is all the easier to do if they appeared to play some part in the downfall of a society which held to them. Maya society in Central America is a case in point: it was toppled rather suddenly, at least from its full splendor, and its end was perhaps hastened by its attachment to the cyclical. The society was governed by a whole series of cycles—a sacred almanac of 260 days, for example, and a year of 365 days divided into 18 months of 20 days each, with a final period of 5 days. These 5 days were in a way outside the year, but they were not a holiday. They were a period of great danger in which people stopped all but essential work, fasted, and were chaste. The almanac and the year were themselves bound into longer cycles of 52 years and those into still longer cycles of 260 years, in which history, which was not our sort of history, was expected to repeat itself. These cycles made up a highly complex calendar which determined where a particular day was on various cycles and what behavior was therefore called for on that day. Each day was divine and, like all other divisions of time, had a god assigned to carry it around its circle. Gods bearing days, months, and years were all members of relay teams coinciding in regular combinations as they marched together through eternity. The priests could always calculate which gods would be together and, hence, what their combined influence would be on human fate on any day, just as astrologers still claim they can divine human destiny from the position of the stars. The priests were, as usual, the managers of time. But they were not always protective of their own people. It may have been partly because the Spanish conquistadors who conquered the Maya in the six-teenth and seventeenth centuries made their appearance on one occasion at the beginning of one of the divine cycles that their invasion was not more contested. It appeared they had been sent.[8]

Cortes did not defeat the cyclical. It lives on in the descendants of the ancient Maya and it lives on in many other places such as Bali.[9] It lives on in our industrial societies, too, if with different manifestations.[10] I am arguing that it would be in accord with the facts of everyday experience, if not with modern stereotypes, to pay more attention to the continued turning of the wheel. Millions of people are still woken by the wheels of an alarm clock, which can be left at the same setting day after day, merely wound up for its journey from twelve back to twelve—unless they have gone digital, in which case that tiny ritual is denied them. The days turn, and the seasons; and the cycles in human behavior, which sometimes correspond to them and sometimes do not, are no less cycles because they are intermittent. Behavior seldom describes a continuous cycle, like the sun or our bodily temperatures, by following a path right around and starting over again with no break; but breaks in behavior do not stop it from being cyclical.

The intermittent cycles can be sequence-locked, being locked into a sequence without being locked into a particular time; and they can also be time-locked, recurring at the same time in a day, a week, a month, a year, or in some other duration.[11]

Life: Cycle or Line?

Is human life a cycle or a line? It is regarded as the former by people who believe in reincarnation. In the past the belief was common: because the dead are not dead the living are not going to die—or, as Edward Leach described it, "death and birth are the same thing . . . birth follows death, just as death follows birth."[12] Even in industrial societies many people still think in this way, especially when they are closer to death than to birth. But their number may be lessening and for those who do not cling to the old comforts the most stubborn fact with which the cyclical view has to contend is death. Death casts its shadow backward, dooming each moment to extinction as finally as it dooms the individual person, a reminder which cannot be long forgotten of what is to come.

Yet life would not be life if it were only a string of endings. The equally stubborn fact with which the linear view has to contend is birth. The individual dies but the collective life to which the individual belongs has its own kind of cyclical immortality. People become parents and grandparents; from the acorn springs the oak; from the egg the butterfly. Grandfathers are always emerging from the womb. In and out of human society renewal is the mode of nature, not just between lives but within lives. If we observe our own habits, however much we would like to believe that each moment is fresh enough to be the first, we may be forced to acknowledge that much of what we do is what we did before and that, when we can make the unconscious conscious, much of what we feel is what we felt before. Van Gogh wrote to his brother, Theo, "Don't let's forget that the little emotions are the great captains of our lives, and that we obey them without knowing it."[13]

People have not changed so much that (if pressed) they cannot recognize something of the Maya in themselves. Modern grandparents are usually pleased at this realization that life is repeating itself, that the events of their childhood, or some of them, are being reproduced again by their own kind, and may even work out better. Most people relish the recurring festivals of the year. The ceremonies that are performed without cavil (if more wryly than they were) can evoke rebirth, revival, and renewal, and have a special quality of timelessness to them: a sense of sacred time as distinct from the profane time of ordinary day-to-day living. People like life going on its circuits. People like the fact that each day is a new beginning, offering some hope of refreshment, but with a kinship between it and all the other days that have rolled themselves away in their lifetimes. In a political system with cycles of elections, those who do not like the president, prime minister, or party in power expect their chance to come around again, and so sustain hope of a new and better day which makes them more ready to accept the disappointments of the present. Democracy's appeal is that of tomorrow, legislators rather than priests holding out the always glittering prospect that the long night will not last forever.

The Passage of Time

I cannot go further without glancing at another, more mysterious matter: our experience of time as something that is always passing, moving or flowing. Does it flow in a line as in a river, or in a circle as around the surface of a globe? A line seems an apt metaphor when the speed at which we are moving (or being moved) in time appears to vary, allowing us to compare the speed at one moment with the speed at another moment. If we were repeating ourselves so that we kept passing in quick order the same spot or moment we would be less certain how the speed changes. As it is, it is more as though we are moving along a road, sometimes accelerating, sometimes decelerating. It seems, for instance, that the pace can be braked into slow motion by certain kinds of time-bending drugs, by boredom, or by danger.[14] The prospect of a car crash can slow the pace right down, as if to give more time for evasive action;[15] and the coming of death may slow it almost to a stop, with a whole life being traversed again in an enormously stretched-out present.

Time also appears to speed up as people move from childhood into adulthood, and even more as they enter old age, distances getting longer as time gets shorter. For a child a day may seem a year; for an old person a year may seem a day. An infant's age is designated in weeks and months, old people's in decades.[16] One researcher estimated with a strange precision that the passage of time for persons twenty and fifty years old was four and six times faster respectively than for a five-year-old child.[17] This perception depends not just on the age of the person but also on whether a time is being experienced or remembered. Memory alters the speed with which time seems to have passed. A time in old age may be boring, relatively empty of new and interesting experiences (because most of what happens has happened before), and so may seem long while it is passing yet short in retrospect.[18] Contrariwise, "a time filled with varied and interesting experiences seems short in passing, but long as we look back."[19] Any experience can simulate life as a whole.

But as soon as memory is in play the linear loses some of its predominance. Along with a sense of motion in a moving present, whether faster or slower than it has been, or simply meandering, we can also perceive a past and a future. With the past and the future the cyclical comes back into its own. This can be brought out by asking the old and ever puzzling question of whether the past and the future exist at all. Those who want to answer no, on the grounds that the future may never arrive, whereas the past is gone forever, are in good company. They are with Aristotle, who said of time: "Some of it has occurred and is not, while some is going to be and is not yet . . . Now what is composed of the non-existent would be thought to be unable to partake of existence."[20]

Saint Augustine gave a different answer—the one I prefer. He acknowledges that past and future do not exist except in the present, but draws a different conclusion.[21] "From what we have said it is abundantly clear that neither the future nor the past exist, and therefore it is not strictly correct to say that there are three times, past, present and future. It might be correct to say that there are three times, a present of past things, a present of present things, and a present of future things. Some such different times do exist in the mind, but nowhere else that I can see. The present of past things is the memory; the present of present things is direct perception; and the present of future things is expectation."[22]

In such a manner, past, present, and future can be distinguished and experienced in the mind as separate entities. They are experienced simultaneously but without fully merging. It seems contradictory to "stretch out" simultaneity only because (as if to mask the athleticism of our thought) we do not allow sufficiently for the speed with which thought operates and are inclined to impute to it, by way of metaphor, the much slower speed—one at a time, one after another—at which actions follow upon each other, or the midway speed of speech. Time passing is the continuous comparison that is made between the future as it slides into the present and the present as it slides away into the past. On this William James adopted much the same position as Saint Augustine. He took over the phrase "the specious present" from E. R. Clay, and pictured it as

no knife-edge, but a saddle-back, with a certain breadth of its own on which we sit perched, and from which we look in two directions into time. The unit of composition of our perception of time is a *duration*, with a bow and a stern, as it were—a rearward- and a forward-looking end. It is only as parts of this *duration-block* that the relation of *succession* of one end to the other is perceived. We do not first feel one end and then feel the other after it, and from the perception of the succession infer an interval of time between, but we seem to feel the interval of time as a whole, with its two ends embedded in it.[23]

According to the Augustinian and Jamesian view the present has to be opened up to make room for every bit of the past which is recalled and every bit of the future which is expected. There is no time other than the here and now for the past to be remembered or for the future to be anticipated. Leonardo and Rembrandt are in the present dancing attendance on the next thousand years. The past and the future cannot be perceived except from inside the present. Conceived in this way, the present is more a dash than a dot, and the stretched simultaneity of the present is what makes possible the sense of movement. This simultaneity can be suggested even by the juxtaposition of the text of a book (which is in the author's present and meant, eventually, to be in some readers') with footnotes which are in both the past and the present. Any observer is in a present which is continually being fed from a recycled past; this is how the cyclical becomes embedded, oscillatingly, in the human present. With apologies to Aristotle but none to Saint Augustine, the past of small people and great people like them, and of small and great societies, is continually being brought back, or recycled, into the moving present, just as the future is "precycled" into the same present. Thus does the leaf flutter, comparisons being made between how it is and how it was, and how it is expected to be. The past in the present allows (indeed requires) the movement between the past and the present to be perceived or imputed. The past also gives some relative permanence to the sense of selfhood of the observer, the I, which is in the present: the I of now can be traced to the I of a moment ago, or years ago, which has been coded into symbols in a memory with some continuity to it. Any present I is in a direct line of descent. We do not wholly reinvent ourselves in each new present.

Although anything remembered from the past is also caught up in time, moving along with the memorizer into the same future, it does not seem quite like that. It seems rather that the memory, surrounded by other memories inside its own moving cocoon, has enough stability to protect itself from being wiped out completely—that the inner world which is created out of transient, consecutive, and irreversible time has a little measure of permanence even while constantly being remade. If all art is a petrifaction of time, designed to trick time out of its victory, so are all memories.[24] We are creatures of time who have yet been given a remarkable capacity, by means of memory and habit, to slow down or even stop time and, where we cannot stop it, to spread it out over millions of years which we think of as having passed, and which, neatly filed in date order, can seem almost as continuously real as the present. The present can be, in Plato's phrase, the "moving likeness of eternity."[25] We know the petrifaction is a kind of illusion. But the illusion, nurtured by habit and memory, is the rock on which all our personalities and all our societies have been built.

The Strung-Out Line

I have been making so much of the cyclical because in the industrial world the stress is the other way. We resemble the conquistadors more than the Maya. We are likely to dismiss people who make much of the cyclical as fatalistic. If we are forced into admitting that of course events recur and do so repeatedly, we can still cling to our preference for the linear scheme by resolutely insisting that the cyclical can be contained within it by adding the cycles to each other. The child is bowling the hoop along the road; each individual cycle can be unrolled, at least inside the mind, and its period measured. A mathematical snail can be made to crawl along a line on the inside of the moving hoop.

Nowadays it is mostly taken for granted, without having to belabor it too much, that time is not like a circle but like a strung-out line, with the past toward one end and the future toward another, with a moving dash in the middle to stand for the present or the now, each new now being a novelty different from any old now, and with segments corresponding to durations, which in our metronomic society can be extended and above all metered. Locke said, "Duration is but, as it were, the length of one straight line extended in infinitum."[26] All things, including ourselves, are supposed to be moving forward along this line from one now to another, or if the metaphor of the arrow of time is used, are being shot forward along it by a brawny archer who has given the arrow enough momentum to keep it flying almost indefinitely. If the future segment of the line is suggested by an arrow, the past is not so much a circle, and certainly not so much a labyrinth, as a more or less straight road along which we are looking back—the standard spatial image being of life's journey from the cradle to the grave, marked out, if we turn around, almost as far as the eye can see, by milestones like leaving school, or turning points like falling in love. Our ancestors diligently hurried along the

same road before us. Sebastian de Grazia stated the modern or-
thodoxy: "Time does not repeat itself, it ticks off in a straight
line, goes from t to t^1 in a continuum, runs in an even flow or in
a stream with graduated steel banks, moves like the assembly line
or the ticker tapes."[27] There is less interest in the past than in the
future, less in permanence than in change, less in the old than in
the new, and altogether less in the cyclical than the linear, as befits
an age which prides itself on being a novel affair. "For various
quite obvious reasons—type of technology, size of population and
organisation, and the nature of the problems engendered by this—
our time is in fact blatantly unique."[28]

The accent is on the future which the linear view finds so
congenial. It allows the future to break away a little from the
cycles of past experience. John Dewey was not only the co-parent
of my old school; he also wrote in *Human Nature and Conduct*
that the human mind is composed of impulse and intelligence as
well as habit.[29] That is one good (if incomplete) way in which a
mind can look at itself. When engaged by impulse and intelligence,
as we all are some of the time and some of us like to think we
are all the time, a spatial metaphor is specially apt: we gaze into
the future, contemplating the many paths ahead along which we
might travel. All the possible paths we have created in our minds
stretch forward rather than circling back, and into their linearity
can be inserted the steps of cause and effect. (I have sometimes
thought of the two sorts of time as Dewey time and Tagore time.)
If we take *this* first step in this direction, it will enable us to take
that step; or *that* first step, yet another. In our private lookout
tower we can exercise whim, play with choice, invite chance in
to join us, weigh up risks, and let our minds run in every direction,
wildly, restrained only by the engine of memory churning away
inside to remind us what happened when impulse carried us away
before, but nearly always in obedience, then and now, to the one
law which impulse, ever curious, will obey: the law of stimulus.[30]
Our minds are never still and our flickering eyes are the windows
to our minds. The modern view stresses the point that to keep
attention alive there has to be change, something new coming
into purview, or an urge to create something new. We may be
prepared to admit that without the continuity given by the cyclic,

there would be no view at all—the eyes, not being trained to see, would be blind.[31] Yet without the prospect of the new, the challenging, which our technological powers have given us confidence enough to embrace, the creature behind the eyes would move into permanent hibernation. Impulse extraordinary is inherent in the animal extraordinary, and, with it, the will and the intelligence to love, think, reflect, desire, invent, imagine, choose, struggle, gamble, calculate, organize.

In this preliminary account of the two dimensions of time I have tried to avoid exaggerating. I hope I have made it apparent that the cyclical depends upon the linear as much as the linear depends upon the cyclical. There is no either/or distinction to be made in the manner in which people behave or in the manner in which they regard events. The two dimensions are best conceived as two often but not always complementary ways of looking at the same thing, two alternative conceptualizations of the same phenomenon which do not exclude each other. It is a matter of degree whether the difference between two occasions, apart from the fact that they are at different moments, is more notable than the similarity. If the similarity predominates, one is back again with the cyclical; if the difference, with the linear. If I stress the cyclical, it is only because we often fail to notice the extent of the similarity between what we do at one time and what we do (or did) at another. People's conscious minds respond to the unique. Especially in an age of mass media obsessed with the new, and the news, we notice the unique more than we notice the recurrent. We fail to realize that although no present event can recur, it yet has similarities with past events and presumably with future events, and that if it did not our sense of order could be shattered.

Biological Analogies

I shall elaborate my main point about the cyclical with the aid of three biological analogies. The first analogy comes from chronobiology. According to the chronobiologists, almost all organisms, including the human, are made up of interlocking cycles or rhythms in the neurochemistry of the body which pulse away in all our organs. Biological rhythms are locked onto the motions of the earth and the moon. All living creatures have to learn astronomy, even in a modern world filled with machines seemingly unhitched to the stars. The rhythms of sleeping and eating, and their accompanying behavioral cycles, are as dictatorial as they ever were.

These imperatives apart, such cycles could operate between people as well as within them. But is this so? How much are social relationships like the organic? The same heavenly bodies are used for timing many daily and annual recurrences in people's activities, as are the humanmade device of the week and a multitude of other cycles. In the factory I visit most people constantly repeat more or less the same actions day after day, week after week, year after year; their gestures are the gestures of a machine. Many other organizations are in the grip of the same metronome. Such regular cycles in behavior secure synchronization and predictability. For a society to exist at all the behavior of its members has to be to some extent mutually predictable; but does this predictability have to be taken as far as it has?

The second analogy is from molecular biology. DNA molecules are self-replicators. By repeatedly making faithful copies of themselves in a cyclical manner (except when subject to a mutation), they maintain stability in the organism of which they are part. Is any similar tendency in behavior responsible for people's following cycles even when their timing does not need to be coordinated with others? People have regular hours for eating, sleeping, shopping, washing, and a hundred other things. Why do they clean their teeth at the same time every day? Why did Immanuel Kant take his daily walk so regularly that his neighbors could set their clocks by him, or my neighbor exercise her long-haired dachshund in the park with a like punctuality? Kant let his neighbors down only twice in his life, once when he heard the news of the fall of the Bastille, the second time because he was too absorbed in reading Rousseau. My neighbor misses only when she goes on holiday.

I shall look at the habits which underlie such cycles in behavior. Veblen considered habit a trained incapacity. But habit is also a trained capacity which lessens the effort of remembering and so frees attention for other things. There is clearly a limit to what

we can attend to. Trying to remember too much at once is stress-
ful; an example is "culture shock," the depression people often
feel on being plunged into an unfamiliar culture. The reason seems
to be that old habits have to be largely relearned—nothing works
as expected, from the water faucets or public telephones to polite
ways of greeting—and extra effort has to be put into continual
attention. Habits can therefore be useful. A habit can also be an
incapacity, when the cycles of behavior it prompts are inappropriate to the situation. The history of every person, every group,
every nation shows that. The question I wish to examine is
whether habit and custom (defined as the habit of a social group)
have a genetic underpinning.

The third analogy is different in kind. I question whether the
cyclical has been given its due in the way modern society is
regarded. But it should be clear that I think the concept of the
linear is as necessary as that of the cyclic, even if there is a shifting
dialectic between them. The central question of the book is about
the nature of the contemporary balance between them. Have we
got it right? To explore this question I introduce the third analogy,
from evolutionary biology. The Darwinian scheme postulated a
great linearity in nature. The evolution it depicted had a clear line
of development running all the way from the amoeba to *Homo
sapiens*. Has there been a similar progression in social development? All societies are changing, if only because of the freedom
which the gift of speech has given to human beings. The concord-
ance between what people do and what they say (or think) is
never complete; the gap has perhaps become greater as the rate
of change has increased, especially in technology. As this has
happened, is society itself in the course of becoming a new species?

The speeding up of technological progress would not have been
achieved unless temporal measurement had been improved. We
do not know what time *is* any more clearly than our ancestors;
most of us are no more sure than Saint Augustine, who has for
sixteen centuries been the patron saint of those puzzled by time:
"I know well enough what it is, provided that nobody asks me;
but if I am asked what it is and try to explain, I am baffled."[32]
We know with some precision only what minute it is, and what
day, what year, what century, what millennium. Methods of mea-

surement such as the Maya or the Balinese now seem idiosyncratic; they have been giving way to a worldwide standardization which to modern minds seems entirely sensible. But one odd outcome is that we seem now to be facing what some people call a "time-famine." The executive speeding in his Rolls-Royce to the next appointment has to seize a car telephone to clinch his deal lest he have no time for it when he gets there or lest the other party by then has left the car phone to which he too may have been momentarily attached.

We cannot do without the cyclic. It is the way life defies the irresistible passage of time, death being a necessary part of life. But although we cannot avoid reproducing the past by means of our marvelous genetic and mimetic systems, we do not have to reproduce all of it. If there is a conflict, as I think there is, between natural rhythms and the rhythms of an industrial society, we do not have to regard the conflict as another fact of nature which we have to put up with. The cycles could be changed. Machinelike cycles in a metronomic society do not have to be the lot of human beings forever.

I have not done more so far than indicate what I mean by my two main terms and some of the respects in which the two are, and are not, complementary. In further bringing out some of the analogies that can be drawn from biology, I am not following some of the adherents of sociobiology (as it is called), who seem to me to reduce social behavior to its biology and make the "socio" little more than an appendage to the "bio." I do not find congenial such views as that "the transition from purely phenomenological to fundamental theory in sociology must await a full, neuronal explanation of the human brain"—although no doubt it would help.[33] I advocate weaving together biology and sociology, but with greater respect for the marvels of society as well as of the organism. It is not a question of first the individual, then society; I hope to demonstrate that society has its own characteristic formations and its own characteristic blends of the two dimensions of time.

I approach the stock puzzle of sociology—the puzzle of social structure—as a time puzzle. Society means nothing unless it possesses a measure of stability over time. Some elements in it must have continuity; although they change they do not change out of all recognition. But how is this stability to be conceived? It could be the static elements, or the synchronic (two of the terms often used in this context), which bestow the stability, with the dynamic or diachronic elements being responsible for change. Comparing social structure to that of a building or to human anatomy has important consequences for sociology. If the realm of sociology is confined to the static or synchronic structure, social change can be left to history. As Anthony Giddens has said: "On the basis of this division, sociologists have been content to leave the succession of events in time to the historians, some of whom as their part of the bargain have been prepared to relinquish the structural properties of social systems to the sociologists. But this kind of separation has no rational justification with the recovery of temporality as integral to social theory: history and sociology become methodologically indistinguishable."[34]

An even more important consequence of the old view is that time is in a sense excluded from sociology, which ends up crippled into taking only snapshots or stills of the social scene rather than movies.[35] But the idea that static and dynamic are contrary has become more and more unconvincing; there are now few sociologists who would make much, if anything, of it. For my part I cannot conceive of any social structure, from ancient Maya to modern Japanese, which is not made up of changes, but with cyclical changes being put in place of the static, and linear changes in place of the dynamic. Society is then less like a structure made of stone than like a bubble tent kept standing by constant puffing. Continuities are preserved by the cyclical changes; even language (one of the more permanent elements in the social structure) is preserved by being recurrently spoken, as the culture as a whole is preserved by being repeatedly reproduced. Cyclical movement, with its constant, lively repetition of what has been, is the nearest response that life provides to the human aspiration for permanence. Permanence is a dance. Durability is not durability unless it is pulsating. The linear changes are the discontinuities within it. Adopt this approach and the old-fashioned distinction between sociology and history is further challenged, if the relationship between the present and the past is also perceived as it has been in this chapter. This is not to say that many historians will agree with the "historical" or evolutionary scheme that I propose.

I shall say more about the cyclic in everyday life and in human culture than about the "trade cycles" and other cycles in the economy, or about views such as Vico's that the history of mankind is an "ideal, eternal history traversed in time by every nation in its rise, growth, decline and fall."[36] My emphasis will be more on the cyclic than the linear, at any rate in the first part of the book, not because I believe that the cyclic could ever (or ever again?) stand alone as that in much modern discourse it has been allotted too humble a place. This tunnel vision has been a characteristic of many students of society, and as a result sight has been obscured of the most elemental of the dynamic forces in the social structure. Perhaps it is uncomfortable to be reminded of the survival of the primordial in human beings who are approaching the second millennium of their recent history, and who have some pride in their civilization and in their rationality. I hope not too uncomfortable, for without recognizing the part of the cyclic in maintaining the continuity which we acknowledge, where we can see it, as social cohesion, we shall not understand society at all. That at any rate is my thesis. I think there is little doubt that modern society (and social science with it, despite the fact that for social science the cyclic is of the essence) has a linear bias to it; and that with this linear bias many natural rhythms have been replaced by artificial ones, a rhythmic society replaced by a metronomic.[37] I shall consider some of the losses consequent upon this revolution of the revolutions.

T W O

Extraterrestrial Timers

> Daffodils,
> That come before the swallow dares, and take
> The winds of March with beauty; violets dim,
> But sweeter than the lids of Juno's eyes
> Or Cytherea's breath
>
> *The Winter's Tale,* 4.4.118–122

Instead of contrasting the cyclical and the linear, I am now going to concentrate strongly on the former, by looking not to the Maya or the Balinese for a model but to chronobiology. A recent explosion of knowledge in this branch of science has forced its way into journals in the neurosciences and many branches of biology, pharmacology, and psychiatry, as well as journals specializing in chronobiology. Some of the outstanding pioneers have been J. Aschoff in Munich, F. Halberg in Minnesota, C. S. Pittendrigh in Stanford, and the late John Mills in Manchester. The new studies have shown that the bodies of human beings, and almost all other organisms, are composed of multiple rhythms, time-locked ones at that. Every bodily process is pulsing to its own beat within the overall beats of the solar system.

Chronobiology has therefore provided a new analogy for sociology, and so given a new look to an old and well-tried model. The living organism has long been one of the favorite analogies for society: people puzzling over the way members make up a whole—how they themselves as members form groups (and the groups form them), groups form institutions, and institutions form societies—have sought guidance from the way the members of the body, organs down to cells, fit together. The individual body has the advantage of being small enough to be thought of both as a whole and as an assembly of parts; society, too big to be easily taken in as a whole, can be better visualized when it is compared to the small. The literature of the social sciences, and writing from long before they were thought of, is therefore full of allusions to the body of society, the social organism, the body politic, the body corporate, the corporation, the body of a church. The analogy has been used to support all sorts of views about human organization.

A stock version of this analogy has been between the anatomy of a body and the structure of a society. This is bound to yield a static view, with the valiant support of the standard sample survey which has been one of sociology's main research instruments. The assumption behind it is that all the people included in the sample, even though they are interviewed over a period of months or years, can be compared with each other regardless of when they were seen. Time is stopped. Mrs. Two O'Clock is treated as though she is the same as Mrs. Eight O'Clock, Mr. January as though he is Mr. March. Revealing as it can be about other matters, the survey procedure offers as little enlightenment about the processes of cyclical change as the dissection of an organism explains its behavior.

The new analogy does not have that same drawback. All the parts are allowed to move continuously, at rest as well as when the skeleton is slumping down its fleshly person behind the steering wheel. Because the new model accords a central place to time, my intention is to approach human society with the assistance of the many allies paraded by the new studies in the form of dancing bees, foraging fiddler crabs, burrowing rats, migrating swallows, and lovelorn fruit flies with vibrating wings. This may look like a digression, but the timers which make the bees dance and the crabs forage with such faithfulness operate over the whole earth, and since we share it with them it would be strange if we were not marching to the same drummer. Like them, we sleep, and do many other things besides, at the command of the sun.

The Bent Thermometer

The basic observation is not new. As long ago as 1845, John Davy, a doctor in the British Army, described the ups and downs in one of the most easily measurable fluctuations, his own temperature.[1] As this was long before anyone could go into a pharmacy and buy a thermometer, Dr. Davy had to make his own. He told the Royal Society about it:

> The thermometer I have employed is a bent one, about 12-1/2 inches long, its bulb about an inch long, and, where widest, half an inch thick; its curvature is about 3-1/2 inches from the bulb, and its stem, to which the scale is attached, nearly at right angles to the bulb, so that when inserted under the tongue, the observer has no difficulty in distinguishing accurately the degrees himself, whether near-sighted or the contrary; in the latter instance using merely a common magnifying glass . . . it is necessary that the thermometer remain in the mouth many minutes . . . a shorter time being required . . . if the mouth had been kept closed for a quarter of an hour previously.[2]

Dr. Davy was conscientious about everything; he took his temperature at intervals every day for eight months to make sure the fluctuations he saw were no fluke. The movements were regular, starting the day with a low and rising to a high in the late afternoon. It is now common knowledge that, if without his persistence, we are all Dr. Davys: we all have temperatures which fluctuate up and down during the day. In sickness, the fluctuation is greater.

In the last few years the same kind of observation has been made about a very wide range of living organisms. The one outstanding common feature is that many of the bodily cycles are based on those of the sun and the moon. The sun is not only the source of our energy; in its relationship to the earth as the earth revolves and orbits it is also the coordinator of the multitude of processes which that energy sustains. Its second role is acknowledged not only by the tiny circle we use for a degree symbol—the last hieroglyph left in our writing[3]—but also by the larger circles that (unless we have taken to digitals) we still use to tell the time. The sun governs with the aid of the moon. The main coordinator is so effective because it is so reliable. According to Genesis 1:16 the moon is "the lesser light," but it, too, is highly reliable. If the two rulers were exactly in phase with each other there would be less advantage in having both of them; fortunately they are not. In whole numbers the moon into the sun will not divide. Neither a year nor a lunar month is made up of an exact number of days, nor a year of months. This syncopation has been as much of an advantage for life as it has been a nightmare for human calendar makers trying to create a linear order. It is only in symbolic form that the heavenly bodies in our solar system can be made to move in a straight line, although the task will be easier if, in the course of elevating linearity, we go on depriving the moon of its role in society until it has none left. The function of the moon as a lantern has, in industrial societies, been almost removed by electric light.

The three distinct markers, day, year, month, and their interaction have been a great stimulus to complexity. Very simple organisms like bacteria are not subject to astronomical timing. Their life is less than a day. Bacteria are about as timeless as life can get. But organisms that live longer need to have more modes of operation if they are to take advantage of and occupy one of the multitude of ecological niches in time that such a complicated natural timing system provides. There would be less complexity on a planet which did not revolve or always showed the same face to the sun, as the moon does to the earth; or on a planet which was not tilted on its axis so that there were no seasonal changes as it orbited the sun.[4] One implication is that extraterrestrial intelligence should be sought first on planets whose environments are made more complex by moving simultaneously in more cycles than one. It may also be the case that life only evolved (or revolved?) at all on our planet (and would on another) because there were cyclical changes in the environment. The original DNA molecule from which all life is descended was probably itself descended from other molecules in the primeval organic soup (to use a time-honored metaphor), but had the unique capacity to make copies of itself. It is nice to think that a molecule may have been partly persuaded to copy itself by the copying of one day after another over many millions of years. If it did occur that way, the sun would have helped to create life not just by its energy but also by its cyclical motion. One sort of motion would have been generated by another.[5]

Daily Cycles

The simplest way to show how, as John Heywood put it, "the world runneth on wheels," is to consider each of the three cycles—

days, months, and years—in turn, and not only as they affect human beings. The first was given the name of circadian rhythm by F. Halberg, to refer to cycles of about a day.[6] (The significance of the "about" will become clear later.)[7] The second set of cycles has been called circalunar or circatidal, and the third set, circannual. The lunar and the annual cycles, along with any others with a period longer than a day, are also called infradian. At the opposite extreme are the many ultradian cycles of less than a day, such as the high-frequency electrical activity in the brain, the heartbeat, and breathing. For the most part I shall concentrate on the three sets of cycles linked to the astronomic timers.

That many organisms hatch, grow, move, respire, photosynthesize, feed, digest, mate, and sleep more at some times of day than others has now been amply recorded. This applies to organisms from the least to the most complex. In one single-celled organism, a much-studied seawater alga (the luminescent, dinoflagellate *Gonyaulax polyedra*), four clear daily rhythms have been observed.[8] Photosynthesis, cell division, background luminescence, and flashing all peak with great regularity at different times during each 24-hour period. Even such a microscopic organism has acute sensitivity to light.

Cell division in humans fluctuates in the same manner. This was not easy to establish, because it required examination of living tissues cut out at different hours of the day. The problem was solved by a doctor in New York in 1939, who realized that circumcision could provide all the tissues needed. By examining boys' foreskins which had been cut off at different times of the day she was able to show that cell division did fluctuate; it was at its peak during the afternoon and early nighttime.[9]

At a more complex level, the same sort of patterns can be found in all the systems which make up the body, such as the nervous system, circulation, digestion, respiration, and excretion. Excretory rhythms, for example, are controlled by the kidney. That people pass less urine at night was first recorded in 1843. Abstinence from drink and food was at first thought sufficient explanation. But it was shown later that there must be more to it: "the rhythm persisted in fasting subjects or those fed identical meals at regular intervals; in those remaining recumbent or continuously active throughout a 24-hour period; and in those deprived of sleep or made to live in constant conditions."[10]

Besides this variation in the volume of urine, components such as sodium, potassium, chloride, and phosphate are also excreted in lesser concentrations at night.[11] This has allowed them to be used as indicators of other physiological rhythms, in studies like the one in which Nurse Bellamy was the subject. If, for instance, the potassium rhythm of night workers inverts, with less excretion by day than night, the probability is that other rhythms have inverted with it.[12]

The production of many hormones varies in the same manner. They are under the control of the hypothalamus gland, which is thereby responsible for the regular fluctuations in temperature which Dr. Davy noticed, and for equally regular changes in metabolic rates. These are most easily measured by oxygen consumption and carbon dioxide output, which in turn reflect the variations in the energy demands of the regular processes in the body.

Excretory and metabolic rhythms are very marked in human beings, and so are variations in performance. Speed at card dealing and multiplication, the speed of response to a light signal while driving, calculation speed in adding numbers, steadiness of the hand, and vigilance in monitoring a radar screen and sonar all vary cyclically.[13] There is a midafternoon dip, whether or not people eat lunch.[14] The siesta practiced in many countries has a good deal of physiological support. Tasks which call on immediate memory can be performed better in the morning than in the afternoon and evening. When things need to be remembered over longer periods the afternoon and evening are best.[15]

There are also variations in the estimation of time. People overestimate a one-minute interval (that is, propose periods longer than a minute) in the morning and in the evening, and underestimate in between.[16] At either end of the day, a period of ten minutes of clock-time may be overestimated to the extent of the subject's thinking that twelve minutes have passed. Time can, of course, only seem to go slow or fast when there is some marker to compare it with. The change in time estimation may possibly have once had some survival value. At dawn and dusk the hunters who were our ancestors were also the hunted.

The Wake-Sleep Cycle

Such fluctuations, marked though they are, fade into insignificance when compared with the ordinary wake-sleep cycle. Not all animals follow it. Domestic cats, for example, have short periods of sleep throughout day and night. But when the sun sets, most diurnal (that is, daytime) animals go to sleep, each according to its nature, with dramatic variations between species. One student of biological rhythms, J. L. Cloudsley-Thompson, says that when the heavenly clocks give their signals, creatures show how many different lessons they have learned from the same astronomy:

> Many species sleep in characteristic poses which we seldom see because the creatures are easily disturbed and react by flight. For example, crayfish and snakes assume rigid and bizarre postures. Fishes often sleep on the bottom of a pond or stream; birds tuck their heads under their wings; and bats hang upside down. Whales, sea-lions and some seals sleep under water, only coming to the surface to breathe. Many small mammals such as the dormouse, potto and some sloths sleep curled up in a ball, while the fox uses his bushy tail as a pillow. Elephants sleep only for a couple of hours at about midnight; a healthy elephant may lie with its trunk coiled up like a rope and its head resting on a pillow of vegetation. Elephants that are sick or upset, however, do not lie down at all, but merely doze for short periods whilst standing up. Giraffe, too, sleep only for a few hours holding their heads erect; occasionally and for brief intervals they may rest them on the ground or on their backs. The need for sleep is probably not only due to fatigue, but also reflects the activity of an animal's biological clock. It is a physiological solution to the problem of how to keep still and quiet in a safe place, without getting bored; dreaming meanwhile may prevent the delicate mechanisms of the brain from deteriorating through nightly disuse. During sleep urine flow decreases and body temperature falls, while hunger and thirst are diminished. These functions are controlled by the circadian clock so that they do not interfere with sleep.[17]

Nocturnal animals mimic in reverse the teeming diurnal creatures and plants that are awake, with most of us, in the daylight. In deserts, the sun is not to be sought but shunned when it is at its least beneficent: the way to escape is to go underground. In the Sahara, at fifty centimeters under the surface, there is scarcely any diurnal variation in temperature. Animals which can burrow that deep can avoid the heat. This is how kangaroo rats, jerboas, gerbils, ground squirrels, and many other small animals manage to survive there. They feed at night when it is cool.

> Avoidance of extreme temperatures by burrowing and the nocturnal habits so characteristic of desert and savanna animals, is associated with the operation of a circadian biological clock, which tells its owner when the time has come for it to emerge from its hole and begin foraging. But for this, in the comfortable darkness of its burrow, the desert animal might never know when night had fallen on the sands above.[18]

Outside deserts, animals threatened by desiccation behave in the same way. Earthworms, for instance: "Water rapidly evaporates from their moist skins if they are exposed to warm dry air."[19] As Darwin noticed, they spend the daytime inactive in their underground burrows, emerging at night to feed, mate, and crawl about on the surface. Amphibians, with their moist skins, are mostly nocturnal. Leaves are also moister at night for night eaters such as antelopes.

The risk from predators at dawn and dusk remains as great for many creatures as it once was for humans. Timing has to be just right: "The adaptive significance of bats emerging from their roosts each evening at the same time in relation to sunset probably lies in the avoidance of predators such as kites and bat-hawks."[20] It is specially dangerous for a nocturnal animal to be about in the light or a diurnal one in the dark.

Lunar Cycles

The moon times another set of cycles. Because the earth rotates in relationship to the moon once every 24.8 hours, high tides occur on average every 12.4 hours, making a large variety of temporal niches within these periodicities for animals of the sea and the littoral. The gravitational pull of the moon causes two tides, on the opposite side of the earth from the moon as well as on the same side. The sun also exerts a pull. The sun's gravity augments the moon's when the sun, moon, and earth are all in line (as they are twice every month, at full and new moon) and counteracts it when the earth is at the 90-degree corner of a right-angled triangle formed with the moon and the sun (as happens every month at the first and last quarters). When the earth, moon, and sun align, every 14.8 days, the double pull and therefore the tides are at their highest (spring tides); when they are in the right-angle position the tides are at their lowest (neap tides). There are two springs and two neaps in a lunar month of 29.6 days, the number of days between new moons. This is also the average period of the human menstrual cycle (not 28 days as often thought) in women who are not on the pill.[21]

Organisms which live in the intertidal zone between sea and shore adjust to two cycles, the 12.4-hour cycle and the 14.8-day cycle from spring to spring or neap to neap. The choice is between being submerged by water and dried out by the sun; organisms survive only by finely tuned adaptation. Those that inhabit the middle tidal zone can get by on the 12.4-hour cycle. Fiddler crabs come out of their burrows to forage at low tide, protecting them-selves from predators by turning darker in color before they emerge when low tide is during the day and paler when it is at night. They are also aware of the difference in the height of tides, being more active at the spring tides. Creatures in the extreme littoral zone above the mean high water mark are under water only at the spring tides; and those at the extreme lower littoral are under water except at the neap tides. They are both governed by the 14.8-day cycle. Grunion fish provide an illustration: at the height of the spring tides they swim up Californian beaches to spawn, leaving the buried eggs to develop in the warm sand. The next spring tide a fortnight later washes them out to hatch in the sea.[22]

Annual Cycles

Annual cycles are as marked in their effects. According to species, trees leaf, plants flower, birds migrate, and animals breed in their proper season—spring in temperate climates or the rainy season

in the tropics—and hibernate in winter or aestivate in the dry
season. But how is the issue determined? It is one thing to observe
the correspondence between the environmental and animal cycles
but another to discover the link between them.

The most common mechanism in the annual cycles is "photoperiodism," a quite different phenomenon from the circannual
rhythmicity which is expressed whatever the light conditions. Photoperiodism is the ability of an organism to measure the length
of the day and time many of its most important functions accordingly. Plants generally can be classified as long-day, short-day, or
indeterminate. There is a critical day length in each of the first
two groups: long-day plants flower when the days are longer than
this critical day-length, whereas short-day plants flower when the
days are shorter. According to their character, flowers find the
slot in the season when they are (in the light of the competition)
most likely to be fertilized by insects and therefore most likely to
reproduce. The critical day-length requirement gives a cue as
accurate as a conductor's baton to ensure that all members of a
population of a given species will bloom at the same time. "This
is essential for efficient cross-pollination, and may well arise from
selective pressures against flowering at the wrong time."[23] As a
result some specialists have to cram their annual fieldwork into a
crucial span of just a week or two. Amateurs (as so many of these
specialists are in Britain, unlike in other countries) may miss a
year's observations if the weather is wrong during the period for
which they book their plant-watching vacations.

Differential cross-pollination not only maintains the unity of
any particular species; it also ensures that each species occupies
a particular niche in the temporal ecology of its habitat. Notions
of ecology are incomplete unless they recognize that time is of the
essence. The grand scheme by which living things maintain the
atmospheric conditions necessary for life on earth—and perhaps
one day will do so on Mars or already do on other revolving
planets, in other solar systems—is itself subject to cyclical change
as is almost everything else living within our general system.[24]
The march of the different kinds of trees through the seasons,
each gaining its leaves and losing them one after the other, rather
than all together, is remarkable partly because it is so regular.

Seasonal triggers can control development even in species whose members live less than a year. For those more long-lived, such as human beings, there must be a counting device at work like that which measures the length of the day, except that it measures the number of years (and divisions of years) during which the ordinary annual changes have occurred. These counters presumably trigger the different phases of growth and decline—or "gate" them, to use a word employed by biologists—while allowing a good deal of variation between one individual and another.

Simultaneous Response

I have been dealing with the three rhythms separately, but in practice they are not separate at all. They are not in the least mutually exclusive. Organisms do not respond to the solar and lunar rhythms separately, one after another; we (meaning all organisms) react to them simultaneously, synthesizing them into one response in a continuing present. A particular phase in a day is at the same time a particular phase in a monthly and an annual cycle. We are not outside the cosmic rhythms but part of all of them. We do not consult our internal clocks intermittently, as if we were wearing daily, monthly, and annual watches on our wrists which we glance at now and then. We consult them continuously and together. They are part of us. Time is not only outside us, it is also inside us, in the form of our internal clocks and their consequences.

The intricacy of these clocks is shown by the way that (along with internal "magnets") some creatures use clocks for celestial navigation. Bees and birds have an astonishing capacity for navigating by the sun, using the same clock that controls their own internal rhythms. This clock guides their behavior in two respects simultaneously. It measures the length of the day from first to last light so precisely that birds migrate on the same day every year, or very nearly so. Once airborne, the birds remain on a steady compass bearing in relation to the sun's position, after allowing for the time of day. When a bird migrates, it notes where the sun is by reference to the sun's azimuth. This is the compass bearing at the point where a line drawn vertically down from the sun would meet the horizon. In itself that information would not be enough for navigation; the time since the sun rose has to be known as well. In order to fly southwest at nine in the morning the bird must keep the sun at 90° to its left. By noon the sun has moved 45° clockwise around the horizon and the bird must now keep it at 45° to its left in order to continue on the same course. It takes constant "readings" so as to fly straight.[25]

Bees are even more wonderful: they not only make the same calculations as birds but communicate them by means of a dance-language. Foraging worker bees use their biological clocks to navigate to a source of food, such as a spray of honeysuckle in a hedgerow. When they return to the hive they pass on the information to their fellows by performing a figure-eight waggle dance in the hive. In the darkness the bees sense the movement of the dancer through their antennae. The angle of the cross in the figure eight as it is danced is the same in relation to the vertical as the angle of flight from the hive relative to the sun at that particular minute of the day. The angle of the dance follows the sun precisely even when the dancing bee does not leave the darkness of the hive for several hours or until the next day. The dancer uses its internal clock to make the necessary adjustments so that it continues to point to the honeysuckle.[26]

Members of human societies cannot make such precise calculations, at least not consciously. But they can still keep track of where they are on three of the main coordinates of time, daily, lunar, and annual, and on many others humanmade, without having sight of the heavenly timers or their light. They can also synthesize the multiple bits of information into one bearing rather as they synthesize cyclical and linear time. At any present moment, at a particular point in the day, month, and year, while taking the one bearing they are aware of its several components and able to separate them out if necessary. But in practice in any present moment, daily, lunar, and annual rhythms become one.

The Free-Running Tendency

How are the rhythms maintained when it is not done by photoperiodism? Sundials cannot tell the time when the sun is obscured.

How do gerbils and other living creatures do it when they cannot see or sense the great light? Their internal clocks evidently cannot be merely slaves of the environment. If they were, biological clocks would be no more than internal replicas of our ordinary mechanical clocks, which we deliberately synchronize with the sun.

The question lends itself to experiment. One of the first and simplest was made in 1729 by a French astronomer, de Mairan. He placed a sensitive heliotrope plant, which like many other plants opens its leaves during the day and folds them at night, in a place where sunlight could not penetrate. Heliotrope means literally "turns toward the sun," but this heliotrope did not need light to turn to. It behaved as though there were light when there was not. It still opened its leaves by day and closed them by night even though light did not signal it to do so.[27] Another experimenter, thirty years later, showed that the leaf movements were not caused by the fall in temperature at night. The plant behaved in all ways as though the sun were doing its customary work when it was not.

Many animals have by now been subjected to experiments like the heliotrope. Bees have been much studied and admired, and not only because of their extraordinary gifts as navigators. In one project which did not call on these talents—the navigation was done by a human pilot—forty bees were trained to feed, under constant conditions, in a bee room in Paris, between 8:15 and 10:15 p.m. Then one night they were put in a box and flown to an identical bee room in New York. Would they begin feeding at New York or Paris time? They kept to the latter.[28] The best interpretation of such experiments is that organisms have internal clocks which operate even when they are deprived of their usual environmental timers.

The next question is about the time that they keep. It turns out that internal clocks have a different daily period from the sun: either more or less than twenty-four hours, never just the same. Much research has been done on this too, for instance, on flying squirrels, which make many flying appearances in the literature. They are ordinarily the most uniform of creatures. When forced to live in constantly dark cages they periodically run on treadmills, just as they do in cages not screened from light. But their bursts of activity do not occur at the same times. When there is only night, the intervals between their periods of activity change. In a particular experiment the period was 23 hours and 58 minutes for one squirrel and 24 hours and 21 minutes for another. The squirrels run more freely, or variably, than when the light of the sun tells them exactly what time it is. This, for all animals, is called "free-running."[29]

Human subjects have been treated like the squirrels, isolated from their ordinary time cues in caves or in bunkers such as those built at the Max Planck Institute near Munich and at the University of Manchester. An English cave researcher, Geoffrey Workman, was one subject. He lived alone in a cave for over 100 days, his only light coming from candles and a small lamp. He spoke once a day over a field telephone to a colleague on the surface and carried a watch in an effort to continue on a 24-hour cycle. But he could not. He neither fell asleep nor woke at his usual time. He went to sleep 40 minutes later than his habitual time on the surface. His sleep-wake cycle—the interval between going to sleep on one night and another—settled at 24.7 hours, that is, out of phase with the sun. When he returned to the surface, according to his calculations he had lost several days from the ordinary calendar. Six cave researchers were observed to stay awake longer than usual, with an average interval of 24 hours and 42 minutes between sleeping.[30]

This finding, that free-running cycles have periods that are circadian, or *about* 24 hours—in humans the periods are nearer 25 hours than 24 hours—does not invalidate the main point I have made about the dominion of the sun in ordinary circumstances. For although the tendency to free-run is always capable of being manifested, internal clocks are ordinarily tied to the 24-hour period. Time cues are all around in the environment, and they act to set and reset the internal clock to the same time as the external environment. M. C. Moore-Ede and his colleagues have (somewhat disrespectfully to the evolutionary marvels which produced it) compared the biological clock to a cheap wristwatch which must be constantly reset because it is liable to run slow. The daily resetting is necessary because the length of day varies continuously. This resetting is done by the sun, and especially by the changes at dawn and dusk; a single pulse of light may be sufficient. Resetting can also be done by other temporal regularities in the environment, such as temperature variations or, for humans, the many cues from any social behavior, such as eating meals, with a regular beat to it. The environmental clock is called a "zeitgeber," or time-giver, by chronobiologists, and the internal clock is said to be "entrained" by it. There is no choice about the entrainment. When a zeitgeber like the sun is in play, it takes command.

The free-running tendency operates within the day as well as between days. After it is set the clock slips toward the free-running period so that, except at the very moment of the entrainment, there is a difference between the time of the environmental clock and the time of the biological clock. Some researchers have put a lot of weight on this difference.[31] They suggest that in a diurnal animal dawn produces a phase advance of, say, 2 hours. The internal clock is thereby advanced 2 hours so that it seems later than it is by solar time and the animal becomes active that much earlier. After that there is a steady movement toward solar time until dusk, when the falling away of light causes a phase delay of, say, 1 hour, so that it does not seem as late as it is by solar time, and the animal continues activity longer. It is a neat way of getting the advantage of daylight saving time and winter time within a single day. (The opposite happens in nocturnal animals.) If the advance is 2 hours and the delay 1 hour, the net effect is to reset the 25-hour free-running period to 24 hours.

The phase differences between environmental and biological clocks could matter a lot. No clock is of much use on its own; the biological clock needs to be related to the solar clock so as to make changes in the environment predictable. To achieve this result the organism needs not just a clock which is synchronized with the sun at dawn and dusk, or whenever it is set, but another clock which shows where the sun is or should be (if obscured) at that point in the day. The free-running clock may be that other clock. The difference between the one and the other would show what "time" it is. The continuous comparison between free-running and solar time could provide the vital information which among other things could account for the skill of the birds and bees in navigation, and which may turn out to be an element in the human sense of time passing.

Where Are the Clocks?

The evidence is in favor of an internal biological clock, but it will not be conclusive until the clock (or clocks) is found. The excitement generated by the search has been similar to that of an astronomer looking for a new star whose presence has been predicted from close observation of the orbits of other heavenly bodies. So far the general view has been that in humans the main clock is a pair of small clustering cell bodies called the suprachiasmatic nuclei (SCN) in the hypothalamus behind the eyes, in the same general location as the "third eye" of Buddhism which is supposed to open in enlightenment. When the SCN is destroyed in other species the general functioning is liable to fail too. The SCN is thought of less as the single biological clock than as the primary driving oscillator, dominant in a hierarchy consisting of many other clocks at every level of organization from the cell to the whole organism.

By far the most striking recent discovery has been the part played by genes, even though the genes encode for a class of proteins that is still poorly understood. This discovery was made in experiments with fruit flies, which, partly because they breed so quickly, are one of the great favorites of geneticists. Some fifteen years ago, Ronald Konopka of Clarkson College of Technology in New York State began the search which led him to isolate a "clock" gene or "per" gene (for periodic) in these flies.[32] He showed that an induced mutation could knock out the fly's clock entirely, destroying the sleep-wake cycle and also the rhythm of the mating song which the flies "sing" by vibrating their wings. Other mutations altered the length of the flies' days and the periods of the male mating songs. Another researcher, Michael Young of Rockefeller University (no relation), said "it seemed that the per mutations were close to the heart of the clock itself."[33] The more per gene product the fly makes, the faster its clocks run.

The per gene is involved in at least two separate clocks in

different parts of the fly's body. The daily rhythms are controlled by a clock in the fly's head and the song rhythms by a clock in the fly's thorax. Young has identified a similar gene in mice, chickens, and humans. There is along this trail of research a great deal more to find out. But at least it now looks as though the different internal clocks may turn out to be genetically controlled; the genes in every cell of every organism may turn out to be the basic timers of the body, which keep it in synchrony with the environment.

Faster and Slower

There are other rhythms besides those linked to astronomy. The body is much more rhythmically inventive—enough to justify fully the use of the word "rhythm," even in its musical sense—than conformity with celestial timers would allow on its own. In its different parts the body imitates the functioning of the heavens, without adopting the same rhythms. Here are some of the main human oscillations with their nominal periods:[34]

bioelectric nervous waves	0.1 sec. per cycle
heartbeat complex	1 sec. per cycle
ventilation	4 sec. per cycle
blood circuit flow	10 sec. per cycle
blood flow oscillations	30 sec. per cycle
metabolic oscillations	100 sec. per cycle
vasomotor oscillations	400 sec. per cycle
fast endocrine oscillations	300–1000 sec. per cycle
gas exchange oscillations	2000 sec. per cycle
metabolic fuel oscillations	5000 sec. per cycle
heat balance oscillations	3 hr. per cycle
circadian rhythms	24 hr. per cycle (approx.)
water cycles	3-1/2 days per cycle
longer-range endocrine rhythms	20–40 days per cycle

It follows that one could still make an analogy with the internal functioning of the body even though there are many cycles in society that do not coincide with the circadian, the circalunar, or the circannual. It could also be that many of the cycles, whatever their periods, have a circadian beat as well. The rates of heartbeat and breathing, for instance, have a circadian pattern superimposed on their basic frequencies, the rates being higher at some times of the day than at others. Perhaps the main circadian rhythm, entrained as it is by the one accurate external timer, the sun, acts as the central regulator for all the others, keeping them in phase with each other. The ups and downs in temperature could be one of the determinants, since temperature is critical for all chemical processes, in or out of the body. It seems more likely, however, that a genetic mechanism underlies all the others.

Homeostatic or Homeorhythmic

Before testing out on society the main analogy of this chapter I ought to discuss the old notion of homeostasis. It has been used to refer to a steady state in an organism. The French physiologist Claude Bernard said thirty years after Dr. Davy that "la fixité du milieu intérieur est la condition de la vie" (constancy in the internal environment is the condition of life). The American physiologist Walter B. Cannon coined the word "homeostasis" much later:

> The constant conditions which are maintained in the body might be termed *equilibria*. That word, however, has come to have a fairly exact meaning as applied to relatively simple physico-chemical states, enclosed systems, where known forces are balanced. The co-ordinated physiological processes which maintain most of the steady states in the organism are so complex and so peculiar to living beings—involving, as they may, the brain and nerves, the heart, lungs, kidneys and spleen, all working co-operatively—that I have suggested a special designation for these states, *homeostasis*. The word does not imply something set and immobile, a stagnation. It means a condition—a condition which may vary, but which is relatively constant.[35]

Cannon went on in his famous book, *The Wisdom of the Body*, to show how each system was monitored and how, by a process of negative feedback, forces were brought into play to restore equilibrium as soon as it was disturbed. A simple example is again provided by body temperature. When it goes up, sweating will tend to bring it down again; when it goes down, shivering will tend to bring it up again. Taken by itself, this notion of homeostasis is of a different set of movements, not cyclical so much as cybernetic, which maintain a steady or static condition. The notion when applied to society is compatible with a generally static view of the social structure.

We now know that this concept of homeostasis will not quite do. According to chronobiology the constant norms are not static but fluctuating. Homeostatic processes restore bodily states not to a fixed point but to a moving norm. A person may sweat to bring down a particular temperature first thing in the morning, because that level is abnormal then, but will not sweat later in the day, when it is not. The tendency toward stabilization is perfectly compatible with rhythmicity.

Cannon deliberately chose "homeo" ("similar") rather than "homo" ("same") to allow for fluctuations and rhythms; my objection is not to that end of the word but to the "static." "Homeostatic" is less misleading than "homostatic" but it is also no longer the right word to use. "Homeorhythmic" is, to my mind, a good deal more apt, for there is nothing static about these processes. Homeorhythmic seems to be a more accurate way of describing a system in which any equilibrium is not static but dynamic and cyclical. The older view can thereby be reconciled with the newer.

Whatever the right word, in light of chronobiology we need a new account of the regulatory systems that tackles one of the chief problems with the homeostatic notion: where are the hundreds of homeostatic engines which would be needed to correct deviations in all processes and so synchronize them with each other? As regulators, they would have to be in constant operation, first to make measurements and then to reverse movements away from what I am now calling the homeorhythmic equilibria. Where would the energy come from to activate them?

The answer may turn out to be that there are no separate "engines," but that the different clocks in the body check each other and, where there is a deviation, adjust the speed of chemical processes so as to bring them back to the usual rhythmic pattern.

The beauty of that would be not only that the economy of life would be shown as being made up of numerous cycles but that the cycles would be mutually dependent and mutually controlling. The whole, like a vast orchestra with many sections, would have but a single conductor, the sun.

In rediscovering and reinforcing what other sun-worshippers knew long ago, science seems to be showing that the sun is pulsing away in all living things, including us. There is a mighty, all-embracing symmetry: from the smallest to the largest systems, and from the smallest to the largest creatures, everything is in motion, not random motion but regular motion in harmony with the sun. The cycles nest inside each other, within the organs, within the bodies, within the environments, within the solar system. We may fear that the world—this kind of world—is going to end; but with every tissue of our beings we recognize that the sun rose today and we are prepared, when night comes, for it to rise again tomorrow.

This account shows that these rhythms, the common rhythms of life, produce not so much uniformity as variety. All forms of life synchronize their own functioning with the sun and the moon, but within these constraints they can display a marvelous range of behavior. Natural evolution has produced a great host of organisms, most of which can manage perfectly well with relatively fixed responses to an environment as highly predictable as ours. But it has also produced human beings with a wide range of responses remarkable for their plasticity. If not for us there would be no such ongoing debate between the cyclic and the linear. Other creatures seem for the most part to manage perfectly well with the cyclic. They can pick up and internalize the pulsing environment of the earth and harmonize their own activity with the changes in it. Gerbils emerge from their desert burrows at dusk. Grunion fish spawn at the height of the highest tides. In us, too, temperatures go up and down, day in, day out, without our being consciously involved.

To portray the processes of life as the chronobiologists do is to underline how much affinity we have with other living things.

Time is of the essence of ecology, as it should be of sociology. We share a temporal environment which declares itself in every moment in the most majestic manner. We share biological clocks which we disregard at our peril. Like all creatures of life, we are also creatures of rhythm in the same large rhythmic dominion.

THREE

Timing of Social Behavior

Time travels in divers paces with divers persons. I'll tell you who Time ambles withal, who Time trots withal, who Time gallops withal, and who he stands still withal.

As You Like It, 3.2.326

I have been describing the body as an array of interlocking (or interflowing) cycles, with their own spheres of partial independence within the solar cycle. Still emphasizing the cyclical rather than the linear, I now question how far society can be likened to that model of the body, with recognizable recurrences of the same kind, whether or not entrained by the same heavenly clocks. I will deal first with daily cycles in work and leisure and then with annual cycles before coming to the cycles of the week, which were not mentioned in the last chapter. Bees and birds do not celebrate Sundays. I shall also consider how far the biological clocks whose effects I have been describing can be held responsible for social cycles as well.

Starting with the daily cycles, it would be surprising if there was no correspondence between organism and society. Even though we are more than our bodies, we are usually zealous to hold onto our occupancy of them. If our bodies follow the circadian patterns, so to some extent must "we." The sleep-wake cycle puts its cast on even the most sophisticated industrial societies. Day after day, we get up in the morning as though it were the most natural thing in the world for us as diurnal animals, which indeed it is; we go to work and return home and eventually, like the dormice and the pottos, we curl up again in the same warm and dark place as before to sleep—longer than our fellow sleepers, the elephants and the giraffes.

Besides this hopping out of and jumping into the revolving bed, society is also shaped by hunger, even if the manner of satisfying it takes so many bizarre forms. In affluent countries no longer are the months before the harvest a time for going on a spartan diet (as in many simpler societies): there is a steady supply of food, and people no longer believe in eating on an empty stomach. Instead, hungry or not, they eat when other people do, according to social rather than physiological decree, but still at more or less regular times. The consumption cycle of one person—often a man—entrains the production cycle of another—often a woman. Those who cook and serve the food must have working cycles, whether in the home or out of it, that fit with one day being much like another, not only at each end but in the middle. The job of housekeeping is a short-cycle affair. Homemakers keep their domestic cycle turning over throughout the day, as shop assistants and clerks selling food keep their supplies turning over; their profitability depends partly on their turnover. To fill the shops and the kitchens other people have to get up in the middle of a Kentish or East Anglian night to deliver their vegetables or milk to London in time for more comfortable consumers rising at more comfortable hours to become more comfortable still. The odd fact is that those who do the most essential work with the highest-frequency cycles—not just growing, buying, and preparing food, but also looking after young children, the old, and invalids—generally have the lowest status. Mary Douglas and Baron Isherwood have pointed out that:

> Their work is to take charge of those services that are counted as necessities. This will include servicing bodily functions, since the living organism needs daily or more frequent care. Bathroom-cleaning, feeding, bed-making, and care of clothes are rightly counted as chores; a chore is essentially nonpostponable, high-frequency. They tend to be ranked as menial tasks, and the goods associated with them, however necessary and intimate, ranked as ordinary stuff, of low value. This association works, even in the very simplest society.[1]

For the greater part of each day, however, after sleep and between eating, people do not seem to be under any circadian dictatorship. When they are apparently free from the double hold of physiology and astronomy, do they still follow cycles of their own or someone else's making? I cannot give a comprehensive answer, but I can supply a few examples, beginning with two organizations I was able to observe for myself: an automobile plant in the north of England and a secondary school in rural Berkshire. Their common feature (in this like every other sophisticated bureaucracy I am aware of) is that much of the coordination they have is achieved not so much by the circadian clock as by the mechanical. I chose the two because they are both complex and different.

Joe Murgatroyd is the works superintendent in the Merseyside factory where I spent some time in order to get a sense of how far the activities of the different people employed were cyclical, within the day and between days. I saw him first on a Friday, which is known by him and everyone else there as POETS Day: Piss Off Early, Tomorrow's Saturday. I do not know how far POETS Day has generally taken the place of "Saint Monday," which persisted for so long in British industry in the nineteenth century—the holiday after a binge that in defiance of their employers people took for themselves, as a kind of compensation for the saints' days they had lost. If it has, Saint Friday would go some way toward explaining why British productivity has not exactly been sparkling.

Not much softened by the human-relations school of management, Joe is the terror of the factory. A short, portly fellow in a well-pressed blue suit, with a mottled face and very bright, penetrating brown eyes, the first thing he does when he arrives in his office, not just on Fridays but every day, is to change into heavy-duty shoes. These can stand up to the lubricating oil whose Eastern-bazaar smell fills the air like wet incense, and which washes out as a white scum from under the duckboards between the machines. Joe is a man of routine. A minute after the last worker has clocked in at 7:30 he is off on his first tour.

His shoes clop-clop like a trotting horse, his short legs pumping up and down to keep the clop-clop so loud that it can be heard above the regular hums and whines and thuds and squeaks from the machines. I almost have to run to keep up with him. As soon as he comes into the No. 4 shed he cocks his ear: he knows where to pounce. He has heard a machine in a distant corner not humming as it should, being as sensitive as Nurse Bellamy was to the breathing of her patients to the breathing of his lathes and borers and milling machines and the plaintive squeaks of the new robots that the management has installed in the hope of keeping up with the Japanese. When he arrives he does not, like her, bend tenderly over the source of the irregularity, but bristles and shouts at him and throws up his hands and calls for the foreman, who comes charging up in an imitation of Joe, looking all the more anxious because of the great effort he is making not to.

Men listen for Joe, and the trotting shoes, almost as alertly as he listens to them. The shoes are his means of filling the factory with his presence. In case his sharp ear has not already detected it, he is called on his beeper whenever a machine breaks down. As soon as he assures himself that the repair is being tackled as it should, and he has learned how long it should take, he considers with one of his foremen how the stoppage can be prevented from interfering with the flow of production—by bringing a reserve machine into operation or transferring men to another line so that they have no excuse for standing around wasting the company's time and money. Almost any manual worker in the factory is supposed to be able to do the job of any other. As in other organizations, Joe and the other managers have put a great deal of effort into dismantling the division of labor for others, as long as their own jobs are not affected. Since one of his chief functions is to improvise means of dealing with breakdowns—and they can occur anywhere in the factory—no one of Joe's days is quite like another, as he stresses several times, his brown eyes smiling at me while he does so. He can speak in such a superior tone because he knows that his job is to keep the factory working to its plan, and according to that plan, most other employees, despite the redivision of labor, have modest jobs which are very much alike from one day to another, rather than proudly different from day to day, as he sees his.

The factory's monthly and daily output is decided by the head office in the light of expected sales—for example, 4,595 rear-axle housings a month or 230 a day if that is the number of vehicles expected to be produced in the main factory. The head office may change daily output at a day's notice. The timing, I was told, has become more and more Japanese since their competition became more intense. Time can be exported, it seems. As in Japan, stocks of anything, including time, must now be minimized. Just In Time, or JIT, is now supposed to be the rule everyone follows, although Joe says that so many other people fall down on what they are expected to do that it often seems more like NIT, Never In Time. He himself is the complete JIT man.

He has to convert these monthly and daily targets into a monthly and daily plan for the deployment of the labor force. If 230 axle housings are needed every day, he knows that it will take five men on the rear-axle line to produce that number, and he can calculate in his head the minute-by-minute speed at which each of the five men is supposed to work to hit the target. He can also consult the factory manual. The practice comes straight from the time study of Frederick W. Taylor.[2] The mechanical clock is used with bizarre precision: the operator on the first process, drill-reaming the spotface, is supposed to take 2.059 minutes, the one who takes over the axle on to his profile miller 1.884 minutes, and so on. On the rear-axle line the length of the cycle (as it is also called) is around about two minutes and each of the five workers has to repeat himself 230 or 240 times every day to fill the quota that Joe has been given and passed on to him.

The workers on that line are timed to fit in with each other and with the other workers who have produced the semifinished castings, which are their material, or will build the finished axles into the trucks at the next stage; the people who are not on production or assembly lines are tuned almost as finely. Harry Masters is a forklift driver, rather isolated in the factory as he is in his life outside. His task is to pick up swarf, digging the oily metal shavings from inside the forests of machines that shed them, rather as a cowman digs out cowshit in a barn. Each machine has to be cleaned out once a day. He never varies the order in which he rolls along from one machine to another, although he could if he wanted. He starts every day at the same place in the same workshop and then goes on to the second, third, and up to the 116th machine. After loading the swarf into skips he drives these around to a bay outside the building where he tips the waste into trucks. Each journey takes about 7 minutes and he makes 58 of them in his usual day. He is so used to it he hardly needs to look at his watch at all. His work cycle *is* his watch.

Almost everything that everyone does in the factory is timed, but that does not mean Joe gets his own way without any demur. Continuous conflict about the pace of work is inevitable because the managers, although relying on custom, are also always trying to overthrow it to their advantage. They rely on a customary level of output accepted by all members of a working group on a common timetable; without that, and the pressure that one worker brings on another to conform to a group norm, managers could be powerless. But they also want to speed up the pace and establish a new and to them more favorable norm. This they may be able to do by switching a faster worker from one position on a line to another where he will have more influence, or by introducing a new machine which, they can argue, calls for a general retiming, or by making some concession being sought by the workers through their union.[3]

The union officials are trying to stretch custom the other way, their way, in order to secure gains for their members—an extra few minutes for a break in midmorning; a day's extra holiday at Christmas; or the upgrading of workers performing a particular task so that they get more money. If they succeed, the gain becomes protected by being regarded as part of the "custom and practice" of the workplace, as the customary term goes. They will not give up anything established as custom and practice without a struggle or a compensating concession. Shop stewards claim to be interpreters of custom. They are, like the managers, trying to make customary law in their favor, and bargaining collectively about it. Both are in agreement about one vital matter, the legitimacy of custom once it has become established, at least until it is changed again, in which case the advantages of the change to each will have to be roughly equalized by a payment or a concession of some kind to the party who is surrendering a customary right or some tiny part of it.

In this factory one conflict that remained unresolved was about the variations between one hour and another. Joe, like managers elsewhere, was for a steady output per hour; the workforce would not conform. They ordinarily achieved the targeted output for each day but they did it by working at a very fast pace for about two hours in the morning after they had settled down, and then using the surplus production as a float to tide them over the last two hours of the shift (for a night shift, the last two hours in the morning) so that they could slacken off then. The electricity load curve for the factory, which reflected the level of output, showed very clearly the day-after-day rhythm they had created. Variation appears to be common wherever people have some control over the pace of their work. In this factory intense activity is succeeded by periods of relative idleness. In many newspaper offices it seems to be the opposite, with people having their "own time," relatively relaxed, earlier in the day, until the timetable imposed by having to get the paper to press begins to exert its pressure. In the factory the victory over the time-study man was as much prized, as when Daniel Bell was writing about the American operators who also practiced "making out" early, and so created for themselves a small area of autonomy within a large, bureaucratic organization.[4]

The School

The preceding example is taken from one type of industry, which is not as typical as it used to be. The proportion of people engaged in manufacturing has been falling over the whole of this century and, of those who are left in it, few are so much paced by machines as the workers who make machines for other people to ride around in. My next illustration, a rural school, is not open to the same objection.

The school caretaker, having cleaned the school the evening before, opens the school gates at 8:15 a.m. The children arrive between 8:30 and 9 (brought by parents who have mostly gone through standard and carefully timed sequences in order to be punctual). Mr. Stone, the head teacher, arrives at 8:40, the teachers from 8:45 on, the van from the school meals service at 11:30, the dinner-ladies from 11:45. Each teacher and each batch of children has to appear at the right points in space at the right points in time. The caretaker has to begin work before the teachers and the teachers often continue with their marking and preparation after the children have left the premises. The engagement into the system of one cycle after another, each in its correct order and at its correct time, is indispensable to the cohesion of the whole. Like any organization, the school stores temporal templates for the repertoire of roles that have to be performed in the right order and at the right times, and enforces the translation of the templates into action.

The ordering could not be accomplished without a pervasive sense of what time it is at almost any minute of the day. Its essential device is very like the bell in a Benedictine monastery. The master clock (as they called it without knowing beforehand that I was going to come nosing around) is in the secretary's office. It has been set on automatic for the last twenty years, without ever being reset, to ring the electric bell in each classroom at the end of each 35-minute cycle, this being the standard length of the teaching or sitting-down period embodied in the school-controlling matrix of the "timetable."[5] Different schools have different "periods," from 30 minutes to an hour; whatever their rule, it normally holds right across the school. The children, except when in double periods, must stop mathematics and start geography at the command of the bell. The headmaster is almost the only person in the school who does not have to break off what he is doing and start up something else at the sound. As chronogeographers would point out, if one class outstayed its allotted time in the physics lab by a few minutes, the next group would be left hanging about outside in the corridor. If this happened often, time-discipline would be lost, and perhaps all discipline whatsoever. The continuing lesson every school teacher imposes for the 15,000 hours or more of their lives the pupils are timetabled is that they have to be "on time." If they internalize that lesson well they are going to feel a little guilty for the rest of their lives whenever they are "late." The most insistent noises they will remember from their school-days are the ticking of the clock and the ringing of the bell.

The school is bound into higher powers by more than the headmaster's clock, which keeps the "Queen's time." It belongs to the larger system of the local education authority, which lays down rules for the starting and finishing times of days, weeks, and terms (all, as in the factory, liable to negotiation with the trade unions, which needless to say represent the adults, not the children); the dates when new staff have to be advertised for, be interviewed, and take up their appointments; the dates by which requests have to be submitted to the office for particular monthly meetings in what is called, with a little huffing and puffing by aficionados who think they are using their own trea-sured jargon, the "cycle" of committees. As the school has to fit into the timetable of the local authority, so at the national level does the local authority have to fit into the annual timetable of the Department of Education and Science for approvals for cap-ital expenditure (insofar as it gives any approvals). The DES is in its turn bound by the general annual timetable of the govern-ment, including, above all, that of the Treasury. Everyone has to abide by the paramount financial cycle whereby money flows to the government through taxes and back from it in expenditure. Money is time.

Regularities in Leisure Time

I have been talking about people at work. In large organizations, they have to be regulated into ensembles, characterized by elab-orate synchronization. But this is not necessarily true of people at leisure. The close timing of the working day in industrial society has produced, in leisure, a new category of time whose essence is that people are supposed not to be so constrained as they are when at work. Even if they must make sure of being on duty in their lookout post when their favorite TV program is screened, in their leisure people are at least not so much under the control of others. To contrast with work, people are said to have "free time." What do they do with their bounty?

Unfortunately, about the regularities and irregularities of leisure activities there is again not yet much evidence. Time-budget studies (as they are called) have been done in many different countries, and have produced some information; but from the standpoint of this chapter the studies nearly all have one serious weakness, that the data are usually for one day only. So no conclusion can ordinarily be drawn about the similarities between one day and another. Sketchy though the accounts are, they do not suggest that freedom brings variation. Ian Cullen and his colleagues from University College, London, made a survey of tenants on a municipal housing estate.[6] Men and women were asked how much was routine of all they did, including their free time. The men said that over 80 percent of their time was devoted to routine activities; for women it was less, but not by much.

The fullest studies so far made anywhere in the world for more than one day have been those done by the BBC, covering a full week from Thursday to Wednesday.[7] From these results it is possible to compare one day with another and see whether, say, between eight and nine in the morning people are doing the same things; or, to allow a little more latitude, see whether they are doing the same things at about the same time in a sequence-locked cycle. The original researchers were not interested in such questions and did not bring out the comparisons between one day and another. But Ian Cullen has done some reanalysis of the same BBC data; his general conclusion, about the extent to which the shape of the diurnal curves describe patterns of repetition, is the same as that which emerged from his own study on the housing estate.[8] By and large, people who get up at 6:30 in the morning on one day apparently do so on the next, and proceed to get their usual cup of tea, read their usual newspaper day after day after day, and take their usual dog out for its usual walk (unless their usual dog takes them), in the same manner and at roughly the same time; and from there they proceed sedately and in good order through their usual mealtimes, their journeys, their shopping, their gardening, their television watching.[9] Within the general regularity there are of course differences. Younger and single people have more diversity in their activities and show more variations between one day and another than the married. Marriage, and particularly young children, make for more routine.[10] Another study made at the Institute of Community Studies showed that social class mattered a good deal as well: the higher the class the wider the range of activities people had.[11] They had more variety in what they did at home and they were more active in sports outside their homes; the probability is (though we did not ask them) that with more diversity went less regularity between one day and another.

There might have been signs of more flexibility had working hours been more sharply reduced. The growing numbers of unemployed people have had their working hours cut drastically or eliminated entirely. But among those in paid work, the higher classes are not working any less. According to the same enquiry, the London region was not unlike the United States, as described by H. L. Wilensky: "With economic growth, the upper strata have possibly lost leisure. Professionals, executives, officials and proprietors have long work-weeks, year-round employment. Their longer vacations and shorter work lives (delayed entry and often earlier retirement) do not offset this edge in working hours."[12] This trend toward less leisure time was most marked in Britain among middle-class women who once had domestic servants. Jonathan Gershuny compared some different time-budget studies.

> In the first period, between 1937 and 1961, middle-class women had an enormous reduction in their leisure time, from 346 minutes per average day to 272 minutes (this corresponds to their even larger increase in domestic work time). And even the very substantial increase in their leisure between 1961 and 1975 leaves them with about a half-hour less leisure per day than they had in 1937.[13]

Those who have shortened the hours they spend in paid work have not necessarily done any less routinized work, if unpaid work is included. Indeed, there has been a good deal of support for the "domestic labor paradox"—that as labor-saving devices have proliferated, not only is it still the case that "women's work is never done" but the amount of domestic work has actually increased. This has not been as true recently.[14] But there seems to be a rather deep-seated tendency to maintain work loads, or indeed any pre-existing pattern.

> If some workers enjoy a decrease in the workweek, there is a tendency to take on second jobs to keep work time constant. If home technology makes it possible for less time to be spent on laundry or cooking, then a more varied wardrobe is purchased or more gourmet meals are prepared. If an automobile makes it possible to cover larger distances in shorter times, residential patterns expand but roughly the same time is expended on trips for various purposes.[15]

The major exception to these homeorhythmic tendencies has been television. In the United States in particular, when people add to their free time they spend most of it in front of their television sets. "It is not clear, in other words, that increased free time will not simply be devoted to the neutral, low involvement and time-killing activity of television viewing rather than to some form of leisure or recreation which can provide a more meaningful, stimulating and integrative function for the individual."[16] Freedom of choice can be exercised to increase routine.

The cycles I have been considering have their geographical aspect as well. Any city in the world shows that. Because cities aggregate the habits of the individuals who live in and near them, the cycles of individual people must inevitably be manifested in the collective. The clocks of the individual become the clocks of the city, and vice versa. Despite the spread of flexible work scheduling, there is in any city still a peaking of traffic flow at the same times, day after day; peaking of demands on eating places day after day at the same hours; peaking in the daily demand for energy for cooking, heating, television, and many other things.[17] The circadian (and noncircadian) chronicle has not yet been compiled in detail. Hägerstrand in the Lund School of chronogeography and his associates and followers[18] have made a famous start in studying these matters; but the main work still has to be done of deciphering not the fingerprints but the time-prints of different cities and comparing them with one another. The ebb and flow of activity could be seen as the breathing of the cities, to use Leonardo da Vinci's image. In his *Order of the Book of Water* he wondered about the tides "whether the flow and ebb are caused by the Moon or the Sun or are the breathing of this celestial machine."[19] One could ask the same about the urban tides. The regularities for the individual are tailored to the larger regularities of the metronomic city.

The Solar Year

The same question arises about the solar year as about the solar day: in what manner has social organization made use of it? In some sections of society the social has to be adapted to the astronomical if for no other reason than that agriculture, horticulture, and animal rearing are still seasonal. Most plants and animals needed for human food are still tuned to solar cycles, and so, therefore, are the people employed in producing and processing them. The tie is less mandatory than it was. World trade has joined the northern and southern hemispheres. More crops can be grown out of season, in the open air or under glass and plastic. Some animals can be bred and managed in a way that would have been impossible in their natural habitat. Tractors allow plowing to be done at times of the year that horses or oxen could not manage, and in the darkness which sends mere animals to their sleep. A reincarnated medieval farmer might be more surprised by the sight and sound of a tractor in the middle of the night with its powerful headlights, its mighty engine, and its great wheels trundling across his earth than by anything else he saw. But he might also notice, with relief, that there is still no way of completely circumventing the short days, the unyielding ground, and the low temperatures of winter. There are still proper seasons for plowing and planting, for weeding and harrowing, and for harvesting. No longer do poor men, as in Piers Plowman, have so grim a "winter time when they suffered much hunger and woe."[20] But the seasons still have their hold.[21]

Although being tied to nature's round is still the lot of the two-thirds of the world's workers employed in agriculture, it is in industrial countries now the exception rather than the rule. Technology has given humanity a degree of independence from the sun. But what is striking is the extent to which the seasons—not the natural ones but the artificial ones which have been superimposed upon the natural—still matter, even to townspeople. Modern society is almost as dependent upon seasonal markers as earlier societies. Not only have most of the old seasonal progressions been preserved in some form, but new ones have been added with such profusion that the old circannual cycles of nature could in the end be outshone by the new artificial ones. In one industry after another producing consumer goods, artificial cycles have been built around nature's and a noncommercial primitivism exploited in order to serve commercial ends. The new does not any longer come just in the spring. In an age of change, spring has become as moveable as Easter, and the clothing industry—the women's, and more and more the men's—has done the moving. Stealing a march on the sun, both spring and winter seasons have been advanced until consumers have been persuaded not only that they need new clothes for spring and summer but that they need them in the depths of winter. The retailers have also been prevailed upon to buy stock ahead of the new fashion cycle, which worsens their cash flow but improves the manufacturers'. The publishing of books has also had as tight and strange a seasonal periodicity imposed upon it. This book must appear either in the spring season, which starts in January, or the fall season, which starts in September. The spring catalogue comes out in the fall and the fall catalogue in the spring.

There is now hardly an industry without its annual exhibition, where new products are launched upon a world not always astonished by yet another fanfare for the new toys, books, yachts, and kitchen sinks. In Britain demand for cars has for a long period been manipulated by selecting August as the month when a new registration letter took effect, which then denoted the birthday year for new cars registered in the following twelve months. The age of all cars could as a result be taken in at a glance, and a quick judgment made (as when looking quizzically at people) whether they looked their age or not, according to their state of preservation. Enough people postponed their purchases until August to produce high sales in a holiday month which would otherwise be bleak for the manufacturers, and old stocks were cleared out before the new models arrived in the fall.

It is not only industry that has been annualized. Politicians have their party conferences on the same days; painters their Royal Academies; punters their Derby; gardeners their Chelsea Show; aircraft connoisseurs their Farnborough; and British musicians their Proms, with the last concert in the series a nationally televized display of patriotism rather like the Superbowl in America. The annual meetings of some organizations are marked by special feasts, by stock ceremonies, or by a rite of passage in the governing body when older members have to think up excuses why they should not retire to make way for youth. All are required by law to have annual accounts and most corporate bodies are like human bodies in having to pay taxes to the government at certain times of the year. To celebrate his own rite of spring (if no one else's) the Chancellor of the Exchequer comes out of 11 Downing Street to present his annual budget as regularly as a fiddler crab comes up from its burrow to feed.

Once an annual occasion is organized, its own little cycle gets whirled up as if by a cosmic regulator into the larger cycle of the solar year. The events that matter almost organize themselves. The dates can be circled years in advance. By being associated with the passage of the year in general, the feelings of solidarity in each organization are enhanced. Durkheim said of representative or commemorative rites that their "sole purpose is to awaken certain ideas and sentiments, to attach the present to the past or the individual to the group."[22]

The Religious Year

The most telling examples of solar cycles, worth more description, come from the religious calendar, and also to a quite marked degree from the sporting calendar and the academic calendar. The main Christian calendar is very much operational. It is a double calendar, deriving from Jewish practice as well as from that of the Roman Empire. Because the Jewish people were originally nomadic their calendar was lunar; the Christian church, which grew up in the Roman Empire with a solar calendar, blended the one with the other.[23] For two months of the year the modern religious calendar is consequently solar, meaning that the seasons of Advent, Christmas, and Epiphany are tied to the same specific days each year, as are saints' days. For the other ten months of the year the calendar is mostly lunar. The seasons of Lent, Easter, Whitsun, and Trinity are readjusted from year to year in relation to the changing lunar position of Easter Day.[24]

The seven seasons of the Christian year still have the same relationship to the life cycle of the inspirer and the same relationship to the church—from Advent, when the birth of Jesus is anticipated; Christmas, when it is celebrated; Epiphany, when Jesus is recognized as the Son of God; Lent, when the death of Jesus is anticipated; Easter, when the resurrection is celebrated; and on to Whitsun, when the founding of the Christian church is commemorated, and Trinity, when the teachings of Jesus are studied. Outside the church, these events are no longer the signals for a general communion although they remain just as important for practicing Christians. For them the Feast of the Annunciation on 25 March, for example, matters as much as ever, to commemorate the annunciation of the good news to the Virgin Mary by the Archangel Gabriel. The date has a special significance, being nine months before the birth of Christ.

The symbolism remains powerful. "The recurring yearly feasts and festivals, forever in a self-contained, circular continuity without beginning and without end, express and refer to the eternal quality of the group and the human species."[25] But many of the rituals that remain have become secularized and individualized, and no doubt lost much of their hold as a result. It is as if for many people the body of the church has gone and all that is left in the ruins is a series of private chapels. A great deal of feeling can still be invested in everyday rites of passage, for example to mark what is sometimes still seen as the solemn transition from day into the eternity of night, when in sleep the ordinary sense of time is so much suspended that in the morning you can never be quite sure whether you have been asleep for 1,001 years or for one night. A professor of economics, rational man *in excelsis,* told me he gets immense satisfaction out of laying the table for breakfast just before he goes to bed every night. Unless he has prepared a ritual welcome for the morrow, he cannot sleep peacefully. As he sets out the cups and saucers he is like a monk at vespers saying his prayers. A psychoanalyst said that the few minutes she spent sitting on her bed each night before undressing, chatting inconsequentially to her husband, were the most timeless and the most treasured moments of her life.

Some of the rites of passage in the life cycle—even marriage— are still matters for collective commemoration. In the annual round, Thanksgiving is a great celebration in the United States, but more generally in the Christian world Christmas, stretching into the New Year, has been left as the ritual occasion of the year for private and public observance of the ending of one year and the rebirth of another, as well as of Christ's birth. Christmas is both end and beginning because it is near the winter solstice. The 25th of December was chosen for its date during the reign of the Emperor Constantine when the solstice occurred on that day, before calendrical changes shifted the solstice to the 21st. The 25th had been the day for the pagan worship of the sun, especially in the ceremonial of Mithraism, which overlapped the early stages of Christianity; the Day of the Sun was appropriated by the new religion just as, much later, May Day was by the labor movement, and then re-appropriated by the Catholic Church in some countries as the "Feast of Saint Joseph the Worker."

Christmas has been more than secularized, it has been thoroughly commercialized. Almost every shop in Christendom, and many outside it, measures out the passage of time up to the birth not with the aid of the Archangel Gabriel but in its own idiosyncratic way, at a period in the retailing calendar when time becomes more than ever money as each day passes. "Only 67 days to Christmas" says the banner on the department store, and although it would be odd to see anyone setting up a crib in a church in early October, it is no longer so at that time to see a young man in his shirt sleeves standing behind a plate glass window dressing a model that one does not have to guess is going to become a Virgin Mary when he has got her clothes on. At Filene's, a large department store in Boston, Massachusetts, a good deal more than a third of the yearly sales are made in the three months running up to Christmas. For shopkeepers, at least, it offers salvation.

Christmas is also more than a commercial jamboree. It is not the main Christian festival—Easter remains that. But in a secular society it is the only annual occasion whose significance as a unifier is remotely like that described by Victor Turner, an anthropologist who studied the characteristics of ritual.

> This is indeed characteristic; at all costs the rubrics should be observed. For it is felt that only by staying within the channels marked out by custom, through which collective action should flow (note here the derivation of the Latin *ritus* from the Indo-European root meaning "flow"), will the peace and harmony typically promised to ritual participants finally be achieved. To complete a ritual, as I have shown, is to overcome cleavages. It is collective man's conquest of himself. For in pursuit of personal and factional ends, men are divided, and in loyalty to their sub-groups, men are set at odds, but before what they conceive to be the eternal or eternally recurrent, these divisions and animosities are annihilated.[26]

Although the festivities of Christmas are mostly confined to the family home, ripples still wash over from it, and back again. Almost every organization has its party, or dozens of parties, before Christmas. The friendliness of people to each other in public places is notable. Higher powers are still invoked.

At Christmas everything has to be done not only just at the right moment but with color and gusto. The Christmas tree has to be dressed in the customary manner; Father Christmas or Santa Claus (with his own mythic history) has to be summoned forth, particularly if the household contains the not-so-long-ago born; ritual foods have to be carefully prepared for the feast. Holly and mistletoe are brought out because they are green and bearing fruit when other trees are dead; at the death of one year which is also the birth of another they symbolize life. This is the season for looking backward and forward. Carrying secularization further, one of my sons was impressed by Christmas as the season when television scheduling was altered. The best films on all channels were kept for Christmas, so that he seemed to be in the middle of a kind of "television summer," with "television winter" not to be feared until May, after a second batch of good programs at Easter. The Christmas *Radio Times* also had a special appeal. Appearing on the Tuesday before Christmas, it gave the television programs for a full two weeks ahead and, best of all, contained a little photograph of the following issue of the *Radio Times* with some of the starring programs for a further two weeks on. He imagined he could see twenty-one whole delightful days into the future. Looking at the little picture, it was as if two mirrors were being held up to each other and he could peep over the top into an endless corridor. Some of the first programs after Christmas are retrospectives. They join ordinary newspapers in presenting their "fascinating records of twelve eventful months" in the traditional reviews of the year. At this period television programs and newspapers also speculate about the prospects for the new year, as it is always called even by those who would not acknowledge themselves confirmed linearists.

The Sporting Calendar

The sporting calendar is also seasonal, with roots in preindustrial history. A friend who is a Yorkshire farmer knew, as a boy before the First World War in an agricultural district, that he was being trained as a beater for the game in the same way his family's youngsters had been trained for centuries. He had to be prepared for August 12th, when grouse shooting began on the moors, for September 1st and death to the partridges, for October 1st and death to the pheasants. After that came the start of the foxhunting season on November 1st. He remembers being promoted at the age of fourteen from a "boy" to a "gun" and being allowed to shoot his first grouse. The dates remain the same for these country sports. But the reduction in the working week and the arrival of the Saturday afternoon holiday have brought other sports more to the fore.

In Britain seasonal and time-locked team games are generally in decline, and sequence-locked sports such as tennis, squash, and running are in the ascendant. But the time-locked ones still have a large following. Two of the most popular—football and cricket—are both highly seasonal, one for winter, one for summer, and both stem from preindustrial times, even though they have since become highly professionalized and commercialized. Another popular sport, horse racing, is similar in both respects. By the 1770s there were over thirty days of racing each year at Newmarket alone, and the pattern there has remained much the same, with Newmarket being the "capital" for racing as Henley is for rowing, Lords in London for cricket, and Wimbledon for tennis.

The sporting calendar contains the same days year after year, using the moveable feast method to stabilize the day of the week rather than the date: the Oxford-Cambridge Boat Race is in March, the Badminton Horse Trials in April, the soccer Cup Final in May, Wimbledon in June and July, and Henley Regatta in July (which always includes the twenty-seventh Sunday of the year), and the Golf Open Championship in July, while cricketers play Test Matches throughout the summer and indeed in the Australian summer, which is the British winter. For other sports, days for the finals have to be fitted in between major events and fitted in they have been, though not always between, with over five hundred regular competitions featured in the calendar—so many that one can legitimately wonder who in Britain has any time for work, even on the first five days of the week.[27]

Most of the competitions go through cycles in which some teams or individuals are "knocked out" as the season unfolds, the general hope always being that there will be a "status bloodbath" and one of the less well-known teams defeat the more renowned. The two last survivors fight it out in the finals. In the most popular games the finals have become national festivals or, in the case of Wimbledon, an international one. There is a spatial as well as a temporal gradient to many of the competitions; if the ground required is not too large and the home ground of either is ruled out because it would give one of them too much of an advantage, many of the finals are held in or near the capital city of London at the climax of their sport's particular season. These annual events are the nearest thing to a replacement for the medieval folk festivals. Television has enhanced the effect and the audience. In the United States this is most notable in the two most prominent events, the World Series for baseball and the Superbowl for football.

In 1985 the nineteenth Superbowl game was at Stanford University in California, between the San Francisco 49ers and the Miami Dolphins. Eighty-five thousand spectators filled the stadium, some of them having paid $1,500 for a $30 ticket. For those not on the spot, parties were held in restaurants, hotels, and public and private places all over California and Florida, complete with large TV screens and with almost the same kind of cheering and excitement as at the stadium. Beyond the party guests, some 60 percent of the U.S. population was estimated to be watching the game on TV, even more than viewed the final episode of "M.A.S.H.," making this Superbowl game different from previous ones.[28]

A majority of Americans, brought together by TV on this occasion as on others, had the opportunity to view the spectacle. It had many traditional elements—the enormous choir of young children singing patriotic songs; the teams of girls in short skirts with pom-poms who danced as it were against each other in the period of mounting excitement before the game; the young men at the height of their physical strength, with each side having its special champion in the quarterback; and the wise old men, the coaches, who as they uttered the traditional platitudes about their prospects were almost as much in the limelight as the players.

Also on the scene were higher powers, if secular ones. President Reagan was more present than if he were physically at Stanford, appearing on a very large TV screen above the stands like an immense superman even larger than the colossus Nero built in Rome. Standing in the White House in front of a tapestry which seemed to depict Jefferson's home, Monticello, he himself was in league with a still higher power—the god of chance. He tossed a coin specially struck for the occasion. He performed this ritual at a signal from Hugh McElhenny, once the king of the halfbacks and now marketing director for Pepsi-Cola and 7-Up in Washington, Alaska, and Hawaii. After the toss the president said to the world, or part of it, in a suitably solemn voice: "May everyone do their best, may there be no injuries, may the best team win." To mounting cheers this is what the players tried to do.

The Academic Calendar

In Britain and much of the United States the third annual round, the university year, begins in September or early October. The stock reason advanced for this schedule is that young people used to be needed for the harvest. In Britain another seasonal influence appears to be more telling. The first university was at Oxford, and Oxford, like other medieval cities, was so plague-ridden in the summer months that lecturers and scholars wanted to get well away from it, into the purity of the countryside, and stay safely there until the danger was past.[29] In any event, the October start has remained a fixture, and the long vacation before it, with the consequence that schools also have their long summer holidays (if not as long as those of the universities) at the same period of the year.

As in other institutions, power is exercised from the top down. By controlling entry in the same manner as medieval guilds used to do and by stipulating what an applicant should know to gain admission, universities not only restrict the numbers of their students but to a large extent control what is taught in the schools. They also control the timing of the school year; so there is no gap between the end of school and the start of university, schools finish their year in the summer, and that is when their ceremonies occur, their plays, their special matches and their speech days, not long after the degree ceremonies at the universities. A further and more drastic consequence of schools' bunching their vacations is that parents have to do so as well, in order to take their children away with them for the great new secular ritual of the time-locked annual holiday. In Britain tourist accommodation, which can be more or less empty or closed down at other times of the year, is packed tight in August. Parents and children are joined by others who have to take their holidays then because their workplaces shut. In some British towns all employers close down together in the same week.[30]

To sum up, the year has become, if anything, more prominent as a marker than it was. The ceremonies which date back to pastoral and agrarian societies have resisted the decline of these societies and been joined by many new occasions suited to our much more differentiated institutions. The annual assemblies of all manner of formal organizations are not vestigial remains from a past when people were more directly dependent upon the natural seasons, but new creations. They are, like the old festivals, occasions for expressing solidarity. A very keen-eyed astronaut circling the globe for several years would notice that below him the changes were almost as regular as in the heavens above him: he would notice not only the regular changes in vegetation but also that the people below go through the same motions on the same days year after year, inching along the roads to Wembley or the Superbowl on the scheduled days in their specks of cars, hoisting their candled trees in the same places every year, filing into the emporia and out again with parcels, crowding into large buildings at just the same times, one year after another. Although each day no longer has its patron saint with his or her prescribed honors, an observer could be pardoned for thinking that some days appear to have secular counterparts. As far as the year is concerned, the main analogy between nature and society is evidently not all that farfetched, with the circannual in some ways almost as marked in society as it is in nature.

The Week

The sun has not been the only master. Humans can create their own cycles without having to rely on the ready-made ones. No other creature has demonstrated so much independence from astronomy. No other creature has the week.

To begin with, the week was invented not to demonstrate autonomy so much as to conceal it. Justification for the seven-day cycle had to be found in the heavens. The Babylonians, who can rank as the first astronomers, found justification in the regular movements of the wandering stars. Eviatar Zerubavel has explained in his excellent treatment that long before Copernicus they could see seven of them: the sun was regarded as but the first of the seven which moved around the Earth.[31] In the order in which they became assimilated into the order of the weekdays, the stars and their English and (for comparison) French derivations are:

Sun—Sunday—Dimanche (from the Latin Dominica)
Moon—Monday—Lundi
Mars—Tuesday (after Tyr, the Nordic version of Mars)—Mardi
Mercury—Wednesday (the day of Woden, from Odin, the
    Nordic version of Mercury)—Mercredi
Jupiter—Thursday (the day of Thor, Donar, or Thunar, the
    Nordic version of Jupiter)—Jeudi
Venus—Friday (the day of Fria, Frigg, the Nordic version of
    Venus)—Vendredi
Saturn—Saturday—Samedi

If the Babylonians had had good enough telescopes to observe Uranus, Neptune, and Pluto, we might perhaps have had ten rather than seven days. The Babylonian attachment to seven was much refined in the Hellenistic world, especially at Alexandria. After Julius Caesar's conquest of Egypt the seven-day week was taken from there to Rome, which had up to then held to eight days, and from Rome it spread further.[32] A later junction was made with the Judeo-Christian tradition of the Sabbath: "Remember the sabbath day, to keep it holy. Six days shalt thou labour, and do all thy work; but the seventh day is the sabbath of the Lord Thy God . . . for in six days the Lord made heaven and earth, the sea, and all that in them is, and rested the seventh day; wherefore the Lord blessed the sabbath day, and hallowed it" (Exodus 20:8–11). Later on, Islam followed the Babylonians in adopting the same length of week, although opting for Friday as the day of worship to distinguish it from the others, as the Christians had chosen Sunday to distinguish themselves from the Jews.

It is easier to explain why there is a seven-day week than why there is a week at all.[33] Some smaller calendrical unit between the lunar month and the day is found in most societies; one reason seems to be that, within the lunar month, meeting days are needed for the exchange of goods and ideas and people between communities. The span of the week for this purpose has varied greatly. Ancient Colombia used to have a three-day market week, ancient Indochina a five-day one, ancient Peru a ten-day week and ancient southern China a twelve-day. Most popular in many parts of modern Africa have been three-, four-, and five-day market weeks. The week acts in the same way as any other calendrical notation: it assists coordination between different people and different activities. As the distance and the relationships between neighboring communities varied, so did the frequency with which markets needed to be held. Standardization onto a seven-day cycle has been made easier, wherever it has happened, by arrangements for storing food over longer periods and for exchange of ideas by other means than face-to-face.

Whatever its ancient origins, the week has been securely established in industrial societies. It still culminates in a day for shopping—Saturday—and this first day of holiday is now part of a longer period, the "weekend." The ordinary working week now ends a day earlier. When it was first reduced, in the nineteenth century, hours were taken off each day. The need to reduce fatigue was decisive. But in this century the working week has been cut not by taking more and more hours off each working day, leaving the six-day working week as it was, but by taking the hours off the Saturday. The half-day holiday on Saturday was followed by the full-day; the precedent having been set, it is probably only a matter of time before there is a Friday afternoon off, followed by a Friday morning off. The length of the day no longer has to be cut for reasons of fatigue; and as people live farther from their employment, the fewer the journeys they have to make to work the better.

As it is, the two days before the start of the new week provide what Sundays did on their own in the past: contrast. In 1986 the Bishop of Birmingham led the opposition in the House of Lords to the British government's bill to legalize shopping on Sundays. He wanted to preserve the character of Sundays as a special day. "It gives space in the midst of a busy week, and it acts as a marker in the rhythm of everyday life." People, he said, "need a rhythm by which to live; a pattern which provides a day of recreation and rest every week, and a day with their families that is different."[34] This argument, along with others, won out later in the House of Commons when the bill was defeated. The case for Sunday trading was, in part, that it is happening anyway, within or without the law. But although Sundays have become more like Saturdays, the two days have maintained their distinctiveness. The marker in the rhythm is still there, and it is indeed more pronounced for industrial than for farm workers. Because the week is not a natural rhythm, many farmers have to work every day, including Saturdays and Sundays, when there is work to be done. Animals do not bow to the week. For many farmers the seasonal rhythms still matter more than the weekly.

Most family members, however, separated during the week by work or schooling, can be together only on weekends. Zerubavel says he knew a boy who used to believe that he must have been born on a weekend, since his mother would have probably been too busy working during the week.[35] At weekends people's schedules have a better chance of overlapping. The weekend is the opposite of the single days Friday, Saturday, and Sunday, which have kept members of different faiths apart.

The weekend also has a shadow. There has for some considerable period been a regular cycle of births that repeats every seven days. In Britain, in Australia, and in the United States, the number of births is lowest on Sundays and reaches a maximum between Tuesdays and Fridays. In Wisconsin in 1972 the number of births on Sundays was 86 percent of the daily average.[36] The variation seems to be due not to any quirk of the biological clock but to obstetric practice, births being hurried on or held back to keep the weekend sacrosanct for the staff, providing a striking example of the manner in which natural rhythms can be interrupted to suit a human timetable, and the infant induced to obey the requirements of the metronomic society it cannot decline to join.[37] The social contract starts in the maternity ward, as it may end in another ward governed by a slightly different cycle. There is an excess of deaths on Mondays from heart attacks and other forms of heart disease.[38] This does not appear to be due wholly to deaths which occurred on a Sunday or Saturday not being certified until the Monday. For men between the ages of fifteen and fifty-nine there is a Monday excess but no Sunday or Saturday deficit. The explanation may lie in the stress of returning to work on a Monday which is no longer Saint Monday.

The Analogy from Chronobiology

Having discussed the main social cycles I can now ask whether the analogy from chronobiology has been borne out. In terms of what people do day after day, year after year, week after week, I think to some large extent it has. Human behavior is obviously not timed as rigidly as that of most other animals, but much of it is more or less regular. The regularities create the social mesh. The same people repeatedly act together, in synchrony. There could be no social institutions unless they did. In these respects the parallel between the biological and the social organism is a close one. Different cycles are present in the organism; different cycles are likewise present in society, each serving to keep the others in place, and to ensure that different individuals engage with each other, and without too much jarring move from one phase in a cycle to another, and from one cycle to another.

The larger whole can be considered as sociorhythmic in the same way the organism is homeorhythmic only if the cycles are superimposed on each other. If the different cycles could be kept apart in practice the main analogy could not be pressed very far. The birds of the last chapter rise up to their migratory flights after the dawn has announced a new day which is narrowly shorter or longer than the day before. In the birds' perceptions, day and year are blended. Nor do people say to themselves, at 4 p.m. I am a daily person; at one minute past, a weekly; and at two minutes past, an annual. They are all of them at once, even if in the course of a day they are sometimes engaged more with their daily cycles and sometimes with others, and even though all cycles, however long or short, always contain traces of the past and premonitions of the future within the shortest and yet the most continuous carrier of all, the moving present. The amalgamation is most noticeable when a person, more institutionalized than the majority, is locked into a number of different cycles, all of which have different frequencies. The process is similar to that described in Fourier's theorem in physics, except that the response is not an averaged response to all the cycles.

Mr. Colclough is a civil servant at the Treasury in Whitehall. At a particular moment on a particular Friday afternoon in 1987 he was sitting at a desk on which were several photographs of a baby and a girl at various ages, with enough persistence between them to make it apparent that they were the same person. He had just had his monthly salary paid into his bank on the due date. He was hurrying to finish some departmental estimates, which were at a particular position in the long pipeline leading to the annual budget, so that he could get away before the rush hour to drive down to the country to spend the half-term weekend with the daughter of the photographs. She for her part was so much looking forward to getting away from the bells of the private school to which he sends her that she was at a fast speed writing her weekly essay so that she would not have to worry about it over the rest of the weekend. The father's mood was serene, meaning by "mood" the pervasive state of mind, the background hum, which acts as a setting for the more pointilliste thoughts which dart about in it. He was serene because he no longer minded "getting behind" as much as he once did: at his age he has reached as high a level in the civil service hierarchy as he is ever likely to get.

We can all play several roles at once, in anticipation and retrospect as well as in outward behavior, each with a different time signature. Mr. Colclough's and his daughter's several different cycles do not just put a particular kind of regularity into their mental sets but belong to a social order as well. They can interact with other people on the same frequencies in the different departments of their lives, and sometimes (as on this weekend) with each other. They are like everyone else in playing a note which is within a beat, a bar, a phrase, and a movement, and also simultaneously within several different beats, bars, phrases, and movements. Society is like Mr. Colclough or the migrating swallow in not pulsing to just one beat after another, but in the case of society to scores of different ones at once. My view is similar to that of Anthony Giddens, that structure is "the mode in which the relation between moment and totality expresses itself in social reproduction."[39]

Society can be looked at horizontally, with the cycles strung out for separate numbered tags to be hung on them for their different frequencies. (I am using this term, as in physics or biology, to denote the number of cycles per unit time, being the inverse of the period.) One cannot stand by and wait a day, or a year, or a lifetime, just to check the periodicities of the cycles. This would be the linear approach. However useful this may be for analysis, it is not how the array is experienced. Experience is a vertical slice, with the different cycles (going way beyond the few I have been describing) being summed up in the mind and in behavior simultaneously; society is that same experience multiplied by all its members. Although the only experience anyone can be reasonably sure of is his or her own, it is part of having the experience that other people seem to be having it too, or if not the same experience a complementary one. The experience of the individual can be consolidated by being shared, or attenuated when it ceases to be, as happens when older people lose a slice of what had been a common past when their friends and relatives die.

We may pride ourselves on the newness of our industrial societies but in fact much of the dovetailing (although not of course all) is of ancient origin. Not only are some of the vital periodicities still those of the ultimate parent of all our societies, the solar system, but the use made of them is of long standing as well. Certain cycles of the day are as old as the species; those of the year, almost as old as society. At various stages new cycles have been grafted onto existing ones without destroying them, as the Christian calendar was grafted onto the Jewish, or Wimbledon spliced into the summer calendar. In that way the new has become legitimized by its association with the old. In such instances the length of the cycle has remained the same; many other machine-prompted and organization-determined cycles have had different frequencies from the ancient cycles in which they were embedded.

Society remains sociorhythmic, although some of the new rhythms are not those of nature but of the organization, controlled more by mechanical than biological clocks.

Rationale

If the analogy has stood up as a matter of description, a much more perplexing question is why it should. In the case of rhythmic physiological behavior the main internal force now appears to be genetic. The clocks that control the timing of biological processes may be genetically programmed. They represent an innate response to the regular changes in the environment which also confers a measure of independence from the environment.

Such biological cycles also give rise to social cycles. I mentioned particularly sleep and hunger; to these could be added sexual cycles and, indeed, the life cycle, which generates so many inescapable social dependencies. At the very least, the specific biological cycles place constraints upon what people can choose to do and be. In addition, the existence of biological clocks is evidenced by the way they seem to assert themselves when they go wrong or are interfered with, as when we suffer from jet lag.

Although daily cycles can have a biological underpinning the week cannot, unless some as yet undiscovered weekly rhythms come to light. The yearly cycle is a more likely candidate as a biological foundation. Annual cycles are evident in some people with mental disorders, and there may be similar cycles at work in other people of which they may not even be aware as long as nothing goes wrong; these may help to secure a happy adherence to religious, academic, and sporting calendars, and hosts of others as well. The year unifies rather than divides. It is rare for yearly recurrences to be condemned as boring. In modern society people are still almost as prone as ever to regulate a good part of their behavior by the earth's orbit, and to reckon annual recurrences as much a part of the order of society as they are of the order of nature. In at least this respect the social still harmonizes perfectly with the astronomical.

One consequence of the salience of annual and weekly cycles (and monthly ones too, although unless a vigorous campaign to save the moon is launched by the ecologists the month may eventually disappear, driven out by the awkwardness of its fit with the metronomic) is that they continue to provide a welcome element of stability in society. Plato, in the *Laws*, written after the *Republic*, was at pains to emphasize the importance of rhythm as a means of social control. What he was seeking in the *Laws* was a less rigid model that the philosopher-king could impose on the *polis* than that described in the *Republic*. Thus, whereas in the *Republic* Plato declared that all change was evil, in the *Laws* he envisaged the possibility of allowing minor changes which would not affect the overall structure of society. These changes he called "rhythm," which he defined as "order in movement." All instructors and pupils should know "the grouping of days into monthly periods, and months into the year in such fashion that the seasons with their sacrifices and feasts may fit into the true natural order and receive their several proper celebrations, and the city be thus kept alive and alert, its gods enjoying their rightful honours and its men advancing in intelligence of these matters."[40] Plato would be satisfied to recognize that if his injunction has not been followed in detail, it has, all unwittingly, been followed in essence.

The day is a very different matter. The weight of modern industrial society has fallen on it; insofar as people have been required to forgo their natural rhythms it is within the day that modern technology and bureaucracy have introduced quite new infradian social cycles which are unknown to nature and certainly do not derive from astronomy. It may be fine to mimic the heartbeat in some circumstances, but not to do so for too long in a single set of gestures with little of the human in them and much of the machine. To my mind, the linear has here taken over too fully from the cyclical, with the consequence that in the modern arena of the day social and biological rhythms are in conflict with each other. The conflict is between what is natural to human beings and the requirements of modern large-scale organization.

Are We All Dancers?

Running contrary to circadian rhythms would not matter so much were the circadian propensity not embedded (as it appears to be) in a more general propensity for rhythmic behavior. Such a general propensity is suggested by the nearly universal appreciation of music and dance in human societies, which may be rooted not just in the biological clocks but in some of the earliest experiences that people have. Adults do not place stethoscopes permanently on their hearts, but in effect this may have been their lot in the past. It seems that in the womb not only can the mother's heartbeats be heard but that the rhythm has a calming effect.[41] A fetus given a rhythmic sound shows a reduction in its heartbeat rate of four to six beats a minute.[42] A normal way to deal with the complaint of a newborn is to pick it up and rock it gently. It is usually possible to calm a baby not in pain by rocking it, and adults too can often be soothed or transported by another kind of "rocking," in music. "Rock-a-bye baby" is standard practice across cultures as it is also for mothers to carry their babies on their left side next to their heart rather than on the right. By carrying them on the left side the mother's right hand is also set free. The same rhythmic propensity may be enhanced by the intense, nonverbal baby talk which is carried on between parents and children and prepares the way for full-fledged language with its many rhythmic elements. Rhythm may indeed be the primary signal which enables one living creature to identify another, as it will be the sign of extraterrestrial life if ever that is discovered; it is also a binding force first between mother and child and then between people more generally.

The sense of sight develops a good deal later than hearing, which is so much more primary that children born deaf are in many ways worse off than children born blind. Perhaps the organization of hearing in the brain weaves time around everything that comes later, before and after the development of rhythmic speech, providing a time envelope to surround all noteworthy perceptions so that each has a time component. The ear lends its time sense to the eye. The ear, associated with emotion, could lend feeling to the reason associated with the eye. The ear could also help mold the propensity for rhythm, music, dance into a human tendency which is all the more powerful for not belonging to a verbal domain divided into different languages. We may all be dancers searching for rhythms we can dance to which have some of the qualities of those described so far, rhythms that are multiple, with several beats being responded to at once, and without any of the frequencies becoming too short and too dominant. If there is such a general propensity for rhythm (and for the wonderfully idiosyncratic variations that it gives rise to) it could reinforce the circadian tendency and, incidentally, accentuate the conflict between the needs of the person and the needs of the organization.

I shall return to the matter of this conflict after I have given more attention to the linear. Up to this point, I have mainly been dealing with the cyclic. The concurrence of the social and the circadian could help to account for some of the daily cycles that have not been much touched by modern organization and, perhaps, for some of the continuing attachment to annual cycles. But the concurrence could not be responsible for the week, nor for many other new cycles without an astronomical tie, nor for the extent to which the great range of new humanmade cycles within the day are observed. Conflict or no conflict, by and large behavior conforms to them. There is another way of stating the problem. In 1893 Emile Durkheim said that "civilisation has imposed upon man monotonous and continuous labor, which implies an absolute regularity in habits."[43] Admittedly, the monotonous and continuous labor is somewhat less than when he was writing, but not by enough to have removed the puzzle. How has civilization succeeded in imposing so much of this kind of labor on the people who keep it going, if only by building and driving the trucks which roll into London from East Anglia in the early hours of every morning?

By and large the chronobiological analogy seems to hold up. The body enacts a series of cyclical processes. They have been well documented in the annals of chronobiology, and will be more so in the future. For want of a thriving chronosociology, the cyclical processes in the body politic have not been so fully recorded. The few examples I have given certainly cannot be decisive. But I hope they are suggestive. Social cohesion appears to take the same dynamic form as physiological cohesion, people's actions becoming mutually predictable and synchronized by their conformity to shared as well as individual cycles. In this respect people have copied nature, in part because they have brought their biological clocks with them into the mechanized society. But biology cannot be wholly responsible by any means for the obedience to humanmade cycles in modern society, especially when they are in direct conflict with natural rhythms. There is no correspondence in nature to the cycles of the car factory. So what keeps Mr. Murgatroyd clop-clopping and his men with their ears pricked?

Habit: The Flywheel of Society

That monster custom, who all sense doth eat,
Of habits devil, is angel yet in this,
That to the use of actions fair and good
He likewise gives a frock or livery
That aptly is put on.

Hamlet, 3.4.161–165

W hy then do people repeat themselves so much? Why do they do more or less the same thing every year at Christmas, or on their own birthdays, or every day as they go about their daily rounds, getting out of bed in the morning, washing, dressing, getting breakfast, reading the paper, opening the mail, walking to the garage or the station, talking to colleagues, telephoning the same people day after day, writing letters which are much like letters written on other days, stopping themselves going into the pub with a twinge of regret, as on other days? It cannot all be due to their biological clocks. People do not settle down to their Christmas dinner by measuring the day's length to the nearest few minutes: they are not birds compelled to fly to the dinner table (or into the oven) at just that precise moment.

Some additional force must be responsible for the regularity of Christmas, although it is one day in Western Europe and another in Eastern Europe, and for its absence in large parts of the world; and for keeping most people head-down at their daily tasks when it is not Christmas. I am in other words looking for a "sociological clock" which is as powerful and omnipresent a synchronizer as the biological clock. I propose that this force is the force of habit and its extension, custom—the tendency we all have, in greater or lesser measure, to do again what we have done before. Habit is as intrinsic to the cyclic (including some of its irregularities) as conscious memory is to the linear. Habit and memory are each means of preserving the past to do service in the present, but in the main for different though complementary ends: the first to ensure continuity, and the second to open the way for change.

Here I want to introduce the second main biological analogy, not from chronobiology but from molecular biology. The special quality of the genes in the DNA molecules is that they are self-replicators. The genes are not only decoded to make particular proteins which continuously form and re-form the cells in the body and are built up into working parts with particular functions within the whole, but also act as templates upon which copies are made of themselves. Each gene reproduces itself and assures itself of a measure of immortality by passing into another carrier. The essential element of this analogy suggests that a tendency to self-replication is as basic to social processes as it is to physiological ones, even though there is fortunately a good deal less fidelity in the copying than there is with genes.

Genetic reproduction is a suitable analogy for my present purpose because it is not precisely tied to external timers like the sun; it is, in my terms, sequence-locked rather than time-locked (though it can be the latter as well). It also has the same merit as my first analogy from chronobiology in showing how a cyclic process can sustain a durable structure. My hope is that in pursuing this analogy, or the two in conjunction, some further light may be thrown on the question of how the cohesion of society is maintained.

From Habit to Habit

The word I am beginning to bandy about is the same word John Dewey needed when he wrote *Human Nature and Conduct* in 1922:

> But we need a word to express that kind of human activity which is influenced by prior activity and in that sense acquired; which contains within itself a certain ordering or systematization of minor elements of action; which is projective, dynamic in quality, ready

for overt manifestation; and which is operative in some subdued subordinate form even when not obviously dominating activity. Habit in its ordinary usage comes nearer to denoting these facts than any other word.[1]

The notion was earlier given a prominent place in sociology. Emile Durkheim, for instance, said that primitive people live by the "force of habit," and medieval people, too: "habit has . . . dominion over people and over things without any counterbalance."[2] Nor had the tendency been by any means superseded in modern societies of his time. Durkheim's general position in 1905–1906 was that "in order to know ourselves well, it is not enough to direct our attention to the superficial portion of our consciousness; for the sentiments, the ideas which come to the surface are not, by far, those which have the most influence on our conduct. What must be reached are the habits . . . these are the real forces which govern us."[3] Max Weber, another of the founding fathers of sociology, used the term "tradition" to the same effect:

> Strictly traditional behavior . . . lies very close to the borderline of what can justifiably be called meaningfully oriented action, and indeed often on the other side. For it is very often a matter of almost automatic reaction to habitual stimuli which guide behaviour in a course which has been repeatedly followed. The great bulk of all everyday action to which people have become habitually accustomed approaches this type. Hence its place in a systematic classification is not merely that of a limiting case because . . . attachment to habitual forms can be upheld with varying degrees of self-consciousness and in a variety of senses.[4]

The need for the word "habit" has become much less in such a context, due I believe not to people being less habit-bound than at the beginning of the century but to the path of development taken by the human sciences.[5] Psychologists of the behaviorist school concentrated on the kind of habits that are "conditioned." Pavlov's famous dogs were conditioned by the rewards of food to salivate when a bell sounded, accompanied by food, and to go on doing so when the bell sounded without any food to follow. The lines of research that stemmed from Pavlov, even when the early approach of Watson was modified and elaborated by Skinner and others, have largely come to a halt, which may be because habit was too narrowly regarded as a response (reinforced or not) to a stimulus, whereas the fact is that far from habit needing a stimulus to generate it, it generates itself unless there is a stimulus to the contrary. The mimetic propensity which can operate in all situations, but which is much less influential in the animals usually experimented on by behaviorist psychologists than it is in humans, was largely left out of account. Most of us are not much wiser about ourselves when we are told that cats can learn to operate a latch to escape from a puzzle box and obtain some fish outside it, and many of the more sophisticated conclusions that came later did not seem to be very revealing either.[6] Generally, far too little scope was left in the scheme for the play of human thought.

The eclipse of behaviorism may have contributed to the spurning of habit over a long period, which is only currently ending; despite all the serious difficulties the notion has, and this book will consequently have, for many intellectuals, it is coming back into fashion, although not under the auspices of the behaviorists. All the same it is worth saying that whether the word is used colloquially or not, it often has a pejorative tone to it. To call anyone a creature of habit is close to being dismissive. I want to challenge such a view by being more positive about habit's virtues: if we were not creatures of habit we would not have survived at all as the kind of creatures we are.

If I am to succeed I need to go back to the time when psychologists like Dewey could write as easily as sociologists like Weber and Durkheim about actions that are (in Anthony Giddens's words) "relatively unmotivated" as being habits.[7] This was before the tendency became so marked to explain behavior as the product almost entirely of reflection and purposive thought.[8] It means skipping back a generation or two to the period before professional specialization had been taken as far as it has. Charles Camic considers that the term "habit," which had long been common to psychology and sociology, was "intentionally expunged from the vocabulary of sociology as American sociologists attempted to establish the autonomy of their discipline by severing its ties with the field of psychology, where (especially in connection with the growth of behaviourism) a restricted notion of habit had come into very widespread usage."[9]

In any event, I think it is now more helpful to revert to the older usage—without dropping the key word, as Pierre Bourdieu has done. He has revived an old practice of Weber and Durkheim by substituting "habitus" for habit. His definition of habitus, "understood as a system of lasting, transposable dispositions which, integrating past experiences, functions at every moment as a *matrix of perceptions, appreciations and actions,*" could as well be a definition of habit, if a cumbersome one.[10] He thinks, wrongly it seems to me, that the word "habit" has to be set aside because it has come to be conceived as a mechanical assembly or performed program. The word is used, at least in the general speech community, much more broadly than that.

Although the statement by Dewey starts with the essential characteristic of habit—the influence of prior activity—many people would now put it differently. Whether influenced or not by the rapid development of molecular biology, recurrence would, I think, now be picked out as the key element. The recurrent or cyclical element in many different sorts of behavior accounts for the use of the word in many different settings—habit of smoking, habit of dress, habit of punctuality, habit of obedience, habit of creativity, habit of mind, even habit of lucid thought, or the habit of revolution. Most later revolutions have been modeled on the French Revolution, or on the Russian, which was itself modeled on the French. I regard most of the cycles entrained in habits as not time-locked but sequence-locked, entrained not by the sun but by the recognition of a situation (which may be no more than the time of day or week or year when a particular sequence of actions would be appropriate) as similar to a situation encountered before. When someone advances toward me with his hand (rather than his fist) outstretched, I usually stretch out my hand to clasp his rather than try to knock him down.

Habits are always being created anew. As the Chinese proverb says, "a habit begins the first time." Habits are generated and locked into place by recurrences so that they become automatic, rather than deliberate. In his *Principles of Psychology* William James gives habit a central place as "the enormous flywheel of society, its most precious conservative agent": "any sequence of mental actions which has been frequently repeated tends to per-petuate itself; so that we find ourselves automatically prompted to *think, feel* or *do* what we have been before accustomed to think, feel or do, under like circumstances, without any con-sciously formed *purpose,* or anticipation of results."[11] For James habit was even more than second nature. He agreed with the Duke of Wellington:

> "Habit a second nature! Habit is ten times nature," the Duke of Wellington is said to have exclaimed ... "There is a story, which is credible enough, though it may not be true, of a practical joker, who, seeing a discharged veteran carrying home his dinner, suddenly called out 'Attention!', whereupon the man instantly brought his hands down, and lost his mutton and potatoes in the gutter. The drill had been thorough, and its effects had become embodied in the man's nervous structure."[12]

This story also illustrates James's statement that "in habit the only command is to start." After that the actions are automatic. "Sec-ond nature" is one term for it, despite the Duke. Another popular term, in a society fancying airplanes for transport, is the phrase "to go on automatic."

There are degrees of automation. When it is complete, there is no need for thinking at all; there may not even be any conscious recognition of the situation that produces the habitual behavior. No proposition is more self-evident than that people take a great deal as self-evident. I just act, without having to reason why. Without thinking about it, I scratch my head, or wink, or open my mouth when I am puzzled, and at a particular corner on my ordinary route to work I go through the routine motions with my arms and feet, all without being aware of it, unless on one occa-sion I put my foot hard on the accelerator instead of the brake or do something else eccentric, in which case I may remember it for life (if there is any left); and I do not do these things because I am so very notably absentminded compared to others but be-cause on such matters everyone is absentminded. We are all to a considerable extent like A. J. Cook, the miners' leader in the

British General Strike of 1926: "Before he gets up he has no idea what he is going to say; when he's on his feet he has no idea what he is saying; and when he sits down he has no idea what he has just said."[13]

Habits can be portrayed in even less flattering terms than that. For Thoreau we are all our own tyrants, like the teamster who drives for Squire Make-a-stir and the Squire himself.

> It is hard to have a Southern overseer; it is worse to have a Northern one; but worst of all when you are the slave-driver of yourself. Talk of a divinity in man! Look at the teamster on the highway, wending to market by day or night; does any divinity stir within him? His highest duty to fodder and water his horses! What is his destiny to him compared with the shipping interests? Does he not drive for Squire Make-a-stir: how godlike, how immortal is he? See how he cowers and sneaks, how vaguely all the day he fears, not being immortal nor divine, but the slave and prisoner of his own opinion of himself, a fame won by his own deeds. Public opinion is a weak tyrant compared with our own private opinion.[14]

More prosaically, I do not have to stir any divinity in me before the product of nine times seven comes to mind or before handing my ticket to the collector as I come out of a London tube station. On many other occasions I may hardly have to pause before I give myself the command and sail effortlessly into my routine— recognize the face and I am smiling and murmuring something about the weather, see the beer and I am reaching for the can opener, open a letter and the standard reply is rattling into my mind. Once fully launched, I may have no more than a vague sense that my mind is working while I am not looking, as though my actions hardly belong to me. Who has not occasionally had to feel whether the toothbrush is damp to make sure that it has just been used? Or found himself or herself getting in the wrong side of the car in a country with a different rule of the road? Motivation, like action, may also be habitual: I take my purpose for granted without necessarily having the means settled. Largely automatic habits take us out of time and into timelessness. The habit is not placed in a duration, and without duration there can be no chain of cause and effect into which one's will can be inserted. Habits also gain force and fixity by joining with other habits to form a personality or a disposition, which is a composite of interlocked habits.

Advantages of Habit

I am seeking to explain cyclical behavior, especially the sequence-locked, but so far I have only produced more examples of it, and given it the new name (or dug out the old name) of habit. I can only ask the reader to bear with me. I have not finished yet. The notion of habit has the advantage that (although many are hidden from view) anyone can recognize it, especially in other people; the word may make this recognition easier because it is in more common currency outside the social sciences than in. But that is only the first step. I have pushed the explanation further along a chain of my own making. If cycles and recurrences can be explained by habit, what can explain habit? Like Saint Augustine, as long as you are not asked, as long as you do not deny you have habits, you know well enough why you act on them; but not if you are asked.

Habits are not usually chosen with any deliberation; they just grow, wild flowers rather than cultivated ones. They would not do this so readily and constantly without a series of overlapping advantages which assure that their growth will not be stopped. I will mention four of them. The first advantage is that habit increases the skill with which actions can be performed. The multiplication table is tiresome to learn but, once it has become habitual, reproducing it is very accurate and very quick. Reading and writing are difficult to acquire in the first place but, once acquired, both can be very efficient. To master the piano is very difficult indeed for the beginner, whose convulsive movements of the body and mangling of the keyboard make it seem impossible that any euphony will ever be achieved; but a few years later the hands may caper over the notes as though to the manner born, and all as the result of unflagging repetition. In such cases an act of will (even if abetted by the cajoling of parents or teachers) can inaugurate a habit which the will does not thereafter have to be engaged in guiding.

The second advantage is that a habit diminishes fatigue. Driving a car, or thinking about existentialism, or speaking a foreign language, or saying our prayers, is tiring the first time it is done, and if a person does not persevere because it is tiring it will always remain so. But persevere, and before too long the same person will be rattling off talk about existentialism while watching a football match or the television, or even driving a car while shouting at his children in the back to be quiet, in the hope that quietness will become as much a habit for them as shouting is for him. If fatigue could not be reduced by such means—or (to put it another way) effort invested now with an immense rate of return in reduced effort in the future—a life of any complexity could be insupportable.

> If an act became no easier after being done several times, if the careful direction of consciousness were necessary to its accomplishment on each occasion, it is evident that the whole activity of a lifetime might be confined to one or two deeds—that no progress could take place in development. A man might be occupied all day in dressing and undressing himself; the attitude of his body would absorb all his attention and energy; the washing of his hands or the fastening of a button would be as difficult to him on each occasion as to the child on its first trial; and he would, furthermore, be completely exhausted by his exertions.[15]

The third advantage is still more significant: a habit not only economizes on the effort put into the humdrum and the foreseen but also spares attention for the unforeseen. A capacity for attention is held in permanent reserve, ready to be mobilized to deal with the unexpected—the truck which appears from nowhere directly in front of one's own car, or the shout for help, or the summons to appear before the boss. Habit, by allowing predictable events or features of an event to be managed with hardly any effort, enables people to concentrate most of their attention on the unpredictable. Habit is necessary to allow this concentration. Without it, people would not be able to cope with the changes in their environments which cannot be reduced to rule; they would be without the adaptability which has enabled them to survive countless threats to their existence. Habits are one of our chief tools for survival.

The fourth advantage—the economizing of memory—in a sense encompasses all the other advantages. To go back to an earlier example, if Mr. Murgatroyd on any morning arrived at work to find he had left his habits behind him, and had only his memory to guide him, he might as well get back into his car and go home. The same is true for the whole workforce. Without their usual collection of habits they would be looking at each other almost as if for the first time, in bewilderment, like a regiment lost in a forest, or an assembly of people with severe Alzheimer's Disease for whom there was nothing in life except today, wondering what on earth to do with themselves while Mr. M. ransacked his memory about the organization of the factory and telephoned the head office for orders. Even if the head office were not stricken by the same disability—and if only one person was left with the capacity for habit he or she would soon rule the corporation, and perhaps the world—it would be a task indeed to translate their orders for axles into routines for everyone in the factory in time to get any made that day. Starting from scratch with only their conscious recollections to guide them, it would be miraculous if he and the other managers decided what exactly should be done, and by whom, in time for anyone else to do any work before the bell for the end of the shift. Without habit, every day would be more than fully absorbed in puzzling about what to do, with none of it available for anything else, until they all decided to give it up and stay at home for good—unless home too was similarly overtaken. It would be too much to have to rely on memory to reinvent the wheel, or the axle, even every year, let alone every day.

The head office might decide to send down for an inspection not another production manager but a psychiatrist. All of us know that some people are like the indecisive imaginary Mr. M., who now does not know whether to put on sandals or boots when he arrives in his office, if he can find where it is. James was severe on such a condition:

> The more of the details of our daily life we can hand over to the effortless custody of automatism, the more our higher powers of mind will be set free for their own proper work. There is no more miserable human being than one in whom nothing is habitual but indecision, and for whom the lighting of every cigar, the drinking

of every cup, the time of rising and going to bed every day, and the beginning of every bit of work, are subjects of express volitional deliberation. Full half the time of such a man goes to the deciding, or regretting, of matters which ought to be so ingrained in him as practically not to exist for his consciousness at all. If there be such daily duties not yet ingrained in any one of my readers, let him begin this very hour to set the matter right.[16]

A century later, we can judge from this that the effortless custodian has changed his stance a little; the man for whom the lighting of every cigar was automatic would be thought the miserable human being now. But that does not reduce the force of the general point: Mr. Murgatroyd and his men have something which is in many circumstances better than memory. They do not have to make room in their consciousness for the past. They do not have to recall what they did; instead they can be guided by the habit of what they did do. Jerome Bruner said about selectivity:

> Selectivity is the rule and a nervous system, in Lord Adrian's phrase, is as much an editorial hierarchy as it is a system for carrying signals. We have learned too that the "arts" of sensing and knowing consist in honoring our highly limited capacity for taking in and processing information. We honor that capacity by learning the methods of compacting vast ranges of experience in economical symbols—concepts, language, metaphor, myth, formulae. The price of failing at this art is either to be trapped in a confined world of experience or to be the victim of an overload of information.[17]

He might have added habits after formulae. A habit is a memory unconsciously edited for action.

Conversion of Memory into Habit

The link between memory and habit is a four-stage process: 1) The very immediate past in the form of a very short-term memory (James called it the "perceived past") is continuously rolling along in the present moment. 2) The immediate past sometimes becomes a somewhat more permanent past (the "conceived past," according to James) by being stored in a longer-term memory where it can be continuously reworked without losing its shape completely.[18] 3) Some of what is stored in memory can be remobilized and recalled into the present moment by an act of will, and some of it spills over into consciousness without being summoned. 4) Some of what is stored in memory can be activated in the present in another way, not by being recalled but by being entrenched in a habit, and the habit (or habitual memory, one might call it, to mark the fact that habit is a harnessing of memory) is not ordinarily subject to recall. The word "habit" lends itself better to being associated with memory in this manner than "role" or even "learning," and allows the process to be conceived of in a way that may be more appropriate for the neuroscientists who are looking for mechanisms in the brain which correspond to what happens, and what appears to happen, in the mind.[19]

Memory is the agency whereby consciousness is converted into habit, or the present into a past with a tenacious duration. The memory does this by acting on recurrences; a particular action or mode of thinking, as it is repeated, can get to the point where nothing extra is added to the recollection by an additional recurrence, whereupon a rather complete transfer may be effected to habit. The acquisition of any skill proceeds like that. What has already been learned has to be recalled as a basis for further learning, but once a certain point in the process is reached, nothing new will be added by an additional recurrence and the learning can be transferred to the automatic. When anyone makes a new acquaintance, say at a weekend meeting, to begin with an effort will have to be made to recall the person's face and to associate it with a name and other facts; as soon as the face is memorized, the associations become habitual so that there does not have to be any conscious recall. The struggle is over, until the next time.

The way the habitual works in such a case is by no means commonplace. The face of a woman from the weekend may remain part of the furniture (or photographic library) of one's mind so that the same face seen again after many years in a bus queue, or even the back of the same person walking along from an angle one had hardly seen her from before, is instantly recognized. She is Helen for me if not for Agamemnon! Unless one had rare gifts of description one would be nonplussed if asked to describe the face or the back, and their reappearance would be required to trigger off the recognition. There need be no conscious recall. The face and the back can spring out of the ancient habit formation as if unbidden, after it has been drilled in by a series of recurrences at the initial weekend. Of course I recognize that there are exceptions; sometimes a nonrecurrent happening can also give rise to a lasting habit, as when a person has such a dramatic experience— a religious or political conversion, falling in love at first sight or understanding quantum mechanics as it were in a flash—that a habit is formed immediately which may last a lifetime. In this case the experience is also likely to remain in the accessible memory. In any relationship the first occasion, or first few occasions, can be particularly significant, setting the form of the relationship which later becomes embedded in habit. The glass has to be designed before it can be mass-produced.

Another way of approaching the subject is via forgetting. Why should forgetting be almost as much an everyday experience as remembering, even for people not yet old? One would hardly think it would be an advantage to the species that so much should be forgotten. But that so much is suggests that amnesia, whether built into habit or not, is functional, no unfortunate lapse but a fortunate one. This forgetting is a particular instance of a more general phenomenon for which I would propose a "law of the disappearing cycle": the more recurrences of any event and the more exact the copies, the less they are noticed and the less they are remembered. This law could apply not only to the failure to spare attention for something which is more or less exactly the same as something which has happened before, but also to the failure to remember it, and hence to the ease with which recurrence is metamorphosed into habit. The failure appears to have a crucially important function. As memory fades, habit takes over. If nature adores cycles, the conscious mind abhors them. The mind adores difference, especially the kind of small variations from moment to moment that keep people alert. This predisposition may also account for the resistance which I and my colleagues who are studying unemployment and early retirement in Greenwich, Tom Schuller and Johnston Birchall, have remarked time and time again in people whom we ask to recall the repetitions in their lives—the resistance sometimes seeming to take the form of boredom so intense that it blocks their memory. The failure on the part of social scientists to do justice to cycles when describing social action may, like the resistance of their informants, be another example of the same reaction. Social scientists are at least as much interested in novelty as their subjects.

The corollary to the law I have just proposed might be termed "the law of the signifying deviation." Recurrences are remembered only when they are not followed. What people remember is the deviation between one occasion and another, such as when learning is taking place. The message and the meaning may well be carried, as in radio, by the modulation, or variations from regularity, in the cycles of people's perennial existence. Without the cycles there would be no modulation; without the modulations there would be no meaning. Harmony without variation would not be music. People are less aware of the regularities, the background hum of their lives, than they are of the departures, the novelties, even though without the background there would be nothing to deviate from. People are much more aware of, and interested in, the persisting changes in society than they are in its persisting continuities.

Forgetting (and what better to forget than regularities?) allows people to attend to changes, and to speculate about what new changes might be on the way. Jorge Luis Borges has a story about a boy, Funes, who remembered absolutely everything from the moment he recovered consciousness after being thrown by a wild horse at the San Francisco ranch near Fray Bentos on the eastern shore of the Uruguay River. He remembered the whole of a book in Latin after one reading although before he had known no Latin. "He remembered the shapes of the clouds in the south at dawn on the 30th of April of 1882, and he could compare them in his recollection with the marbled grain in the design of a leatherbound book which he had seen only once, and with the lines in the spray which an oar raised in the Rio Negro on the eve of the battle of the Quebracho." In other words Funes remembered too much. He could only lie in his cot, remembering. He was hardly capable of thought.[20]

In focusing on forgetting, and its function as the background from which conscious meaning is plucked, I am adopting the point of view of psychoanalysis but generalizing further from it. Freud considered habits uniquely important, especially the kind which are acquired early in life, and would therefore be tenacious on that account alone, and which also are acquired in such a manner that they are particularly pervasive. The past has such a long arm it can tyrannize the present. Once a manner of behavior has been incorporated into the personality as a defense against the anxiety produced by conflict, there is a "repetition compulsion" to go on doing the same thing even when it is not appropriate. Defenses have to be strong in infancy because infants cannot wait. They have not learned that time is a flux. For them time is only the now; gratification cannot be postponed.[21]

According to Freud the timelessness of infancy is extended into the timelessness of the id, in the unconscious mind. "The processes of the system [of the unconscious] are timeless; i.e., they are not ordered temporally, are not altered by the passage of time; they have no reference to time at all."[22] Elsewhere, Freud put the point in this way:

> We perceive with surprise an exception to the philosophical theorem that space and time are necessary forms of our mental acts. There is nothing in the id that corresponds to the idea of time; there is no recognition of the passage of time, and—a thing that is most remarkable and awaits consideration in philosophical thought—no alteration in its mental processes is produced by the passage of time. Wishful impulses which have never passed beyond the id, but impressions, too, which have been sunk into the id by repression, are virtually immortal; after the passage of decades they behave as though they had just occurred. They can only be recognised as belonging to the past, can only lose their importance, be deprived of their cathexis of energy, when they have been made conscious by the work of analysis.[23]

The timelessness of the unconscious is reflected in the timeless or out-of-time quality of dreams.

The grip of certain experiences which are locked into a timeless past cannot be loosened, and the resulting habits can (but do not need to) have a pathological character to them. If I am right, such infantile habits can be regarded as but a special case of a much more general phenomenon. The tendency of childhood is also the tendency of adulthood, and is a necessity for life (although it can at any age be pathological if so securely fixed, or fixated, that it cannot be modified in the light of reason and experience). The intensive study of the special case should be helpful in the further study of the more general phenomenon by psychoanalysts in conjunction with psychologists of other sorts and with sociologists.

The Key to the Lock

Automation is by no means all there is to habits. Unless they have become so deeply ingrained that they are completely automatic, the conscious mind has a vital reflective and monitoring role. A good analogy is an automatic pilot in an aircraft. The human pilot can hand over to the automatic pilot while he or she eats lunch. The robot will fly the plane perfectly when all is going well—indeed (as with most habits) it will in these circumstances do so with more skill, less fatigue, and more accuracy than the human pilot using all the resources of his or her conscious mind. For one thing there is less danger from a loss of memory. But the human pilot remains in final control to deal with situations which no programmer could foresee, deciding when it is appropriate to go on automatic, whether the automatic is functioning as it should, and when it should be disengaged. Many machine-monitoring roles are similar. So it is with habit. When not driven by irresistible compulsions, a person to some extent decides which habits to call up, depending on the situation. We do not wind up our tennis serve if the game is ping-pong. We also observe whether a habit is appropriate, and, if it is not, may sometimes be able to modify it or even break away completely from it. The reflective mind can be in control of habits, subject to the complete personality. People differ, of course. Some people are clearly creatures of their habits and make them into addictions or obsessions: some lock their habits into place and are expert at losing their key. They may not be able to find it even with the aid of a locksmith in the form of a psychoanalyst, whereas others behave as though, without a generous supply of keys which they delight in turning in the well-oiled lock, they will be as lost.

Many habits have some flexibility, behaving more like a grammar than a particular sentence. They conform to rules but within them can be comfortably varied. Someone can play the piano without always having to play Chopin (just usually) or drive a car without always having to go through Trafalgar Square. In practice any drawn-out sequence of behavior is a matter of almost continuous interaction between reason and habit, in which reason is the executive who leaves most of the routine tasks to his assistant. This applies even to the most basic habit of all, the essential social habit, the way people become habits for each other, with the habits occasioning grief when they are broken. No marriage, no family, no deep friendship, no factory, no laboratory would hold up without that.

Writing is another example. The way I hold my pen and form my letters is entirely habitual. I could hardly alter my style of handwriting, however much I tried. It has become a part of me that I and others (unfortunately for them) have to live with, although they may with much effort acquire the reciprocal habit of being able to read it. There are also other habits involved—my writing in English and my style, or lack of it, which could be as much something that is tenth nature to me as my handwriting. But it is not quite automatic writing, like automatic sleepwalking; the content of what I write is not just habitual. Habit does enter in; I think in certain ways that are customary for me and for others with a similar background in a particular part of the world in the twentieth century, laid down from my own past experience and theirs, but I am not always writing exactly the same thing as they or even as I did before. I can harness the collection of habits to some new use, and attempt to vary the style or even the handwriting by an iota, in order to fulfill my intention. My intentions seem to be in command even though they could not be effected without the collection of habits that go into my expressing them in writing. It would be even more difficult to communicate if I had no pen or typewriter and had to dictate everything I wanted to say, and more difficult still to use Esperanto or another language foreign to me.

Likewise a parliament. It is governed by rules which have become habits; this happens with many such rules, norms, and values as they are internalized. If a member is going to raise an issue he has to do it according to the rules, with so many days' notice, perhaps with backing from a certain number of other members, in a set form, delivered to a certain office; and to get agreement about an issue, further sets of formal and informal rules—or habits—have to be observed. The bevy of rules is the grammar of the place: you have to use them. But once again an immense number of different messages can be spelled out within and through the use of the grammar.

Another, much more fundamental example is the growth of a child. Growth consists, among other things, of the acquisition of habits. They do not just unfold under the prompting of biological triggers; they are prompted as well by impulse, which is also a faculty of the mind. Dewey's illustration is of a child learning to walk.

> All habit involves mechanisation. Habit is impossible without setting up a mechanism of action, physiologically engrained, which operates "spontaneously," automatically, whenever the cue is given. But mechanisation is not of necessity all there is to habit. Consider the conditions under which the first serviceable abilities of life are formed. When a child begins to walk he acutely observes, he intently and intensely experiments . . . What others do, the assistance they give, the models they set, operate not as limitations but as encouragements to his own acts, reinforcements of personal perception and endeavour.[24]

Thought (which must include the irrational as well as the rational) does not act only as a check on habit. That would be to give it a subordinate role, when it is at least to some extent independent. In the form of impulse made conscious, it pushes the infant into new activities as the infant grows. Impulse is prompted even in adults by curiosity and continuous exploration.

> Impulse determines the direction of movement . . . Our attention in short is always directed forward to bring to notice something which

is imminent but which as yet escapes us. It is, in logical language, the movement into the unknown, not into the immense inane of the unknown at large, but into that special unknown which when it is hit upon restores an ordered, unified action ... With habit alone there is a machine-like repetition, a duplicating recurrence of old acts. With conflict of habits and release of impulse there is conscious search.[25]

The conscious search is guided by the other faculty, intelligence, which forms a range of intentions that subordinate habits to their command and choose between possibilities for realizing the intentions. Both impulse and intention belong to the linear and continually interact with habits, which belong to the cyclic. It would be strange, indeed, if the entente were always cordial.

If such interplay between reason and habit were not possible, there would be no second system of order to add to that of the genetic. We should all be bound into repeating the past, except where there were random errors in copying habits from one person to another, in the same way we are bound by our genetic systems. The habit system would in its effects be indistinguishable from the genetic. As it is, the two are distinct; people can sometimes choose their habits. The person on the horse can sometimes ride the horse rather than the horse riding him.

My examples, apart from a parliament, are of individuals acting with some deliberation. Individuals do disengage automatic pilots, do play the piano and behave in other ways with some deliberation, or, to put it differently, with some exercise of their freedom to choose. Because habits, if traps, can sometimes be escaped, you can to some extent select those you are going to be trapped into or stay trapped into.

If habits remained under conscious control, however, their advantage would be lost. If the linear could always prevail over the cyclic, if the key could always be put into the lock and turned easily, no benefit would be gained. The mind would have to remain conscious of everything that was going on, and be ready at any moment to jump in with the key. The effectiveness of habit depends on a partial suspension of consciousness and reason, of a handing over of control to another sector of the mind. I do not believe that reason could suppress itself so continuously by deliberation, or, if it did, one would need just such a mechanism as I have been postulating. A habit is an expression of one of the most fundamental cyclical devices of the mind, employed to take experience out of the flow of time; the past, with a measure of autonomy, invades and takes over the present, but on the whole benignly and usefully, not pathologically. Impenetrability is habit's strength and its weakness. For a habit cannot necessarily be modified or abandoned just because it is no longer appropriate, particularly because we may have no means of knowing if it has ceased to be so. All of us are old soldiers liable to drop our meat and potatoes in the gutter. If habit makes us better off in one way, it makes us worse off in another. The first strategic choice for any person, or any society, is between that which should belong to reason of a reflective kind and that which should belong to habit.

Often enough another choice has to be made. For the further fact about habit, and custom too, is that they are ordinarily not in conflict with reason but supported by it. When habit is unconscious, banished temporarily or permanently at the behest of the "law of the disappearing cycle," it acts as an effective driving force. But when it becomes conscious it is ordinarily almost as effective. Reason of a kind leaps to the defense of habit or custom with the usually decisive statement that I, or we, have always done it that way. Actions, when repeated, create not only their own motivations but their own supporting reasons. The burden of proof is then on the challenger to show that it is no longer sensible to do as has been done. Even in these circumstances the most telling argument may be that what is being done currently is not in accord with how things used to be done: the source of innovation is found in reconstructing past practice, usually with a novel twist which suits the challenger. The past is always being invoked to right the balance of the present. So the strategic choice is not so much between habit and reason but between the new (often masquerading as the old) and the old (sometimes masquerading as the new), or between the relatively less and the relatively more permanent, or, one would better say, between two opposite strategies for pursuing durability.

Force of Custom

I now take up again the question posed at the beginning of the chapter, about the force which maintains the reproductive structure of society. I have argued that people have a propensity to repeat themselves cyclically, which accounts in part for the readiness with which the repetitive solar cycles have got so securely embedded in human behavior. This claim would be more convincing if a word other than "habit" were used as well. Habits are characteristic of individuals, whereas my examples of daily, weekly, and annual cycles have mostly been of collective behavior, many people joining together to do the same thing repeatedly at the same time on different occasions. In such instances a word is needed to refer to a shared habit, and there is an equally homely one: custom. Custom can be defined as the habit of a group, a social habit. The force I regard as responsible for cycles in society is custom.

It would also be almost as good to talk about tradition, another word which conveys more explicitly the prescriptiveness of social habits—the implication that what is traditional deserves to be adhered to—and also suggests custom's source. The etymology is a clue. The Latin "tradere" meant primarily "to hand over," but the meaning of "to hand down" has become dominant. The title of tradition can be applied to an idea or a practice which has been repeated in a particular way, by being handed down, and which deserves respect on that ground alone. This is most explicit when the handing down is from parent to child: because it has been handed down from a height and over time, from one to another of two closely conjoined people, the idea or practice deserves to be accepted or followed. "Will you mock at an ancient tradition, begun upon an honorable respect . . . ?"[26] Tradition is usually reserved for a custom which is ancient, or at least of long standing. Customs can spring up as quickly as habits. A temporary society which lasts no more than a day, at a conference, on a bus, or in a boat, can establish temporary customs, often to the delight of the people involved. Adults can get as much pleasure as children out of creating short-term customs which have an element of make-believe to them, even out of the invention or use of a jargon which becomes a tiny private language of the day. But such customs, or others which have already had the make-believe squeezed out of them, would not usually be called traditions until they had gained an extra measure of inertia from having been repeated many times over a drawn-out period. As the present disappears into a past from which it can be made to reappear, so a custom that has entered the past has to wait until it re-emerges as a tradition. Because custom is the more inclusive term, I shall stick with it as the word for social habit.

Neither word by itself explains where the authority of social habits comes from. Perhaps the most celebrated statement on this was made by David Hume, for whom custom and habit were interchangeable. When two objects or actions are constantly conjoined, the perception of one does make us expect the other, and we believe there is a necessary connection between the two. But this is not so. To be so, nature would have to be so uniform that once A had been followed by B a very large number of times, it would be necessary for it to be followed, invariably, the next time. But nature is not uniform. However many times B follows A, there is no necessity for it to happen the next time. Experience alone does not enable us to prove that A is the cause of B.

Why then do we find it so easy to infer the existence of one object, event, or action from the appearance of another that has constantly accompanied it? Hume states the principle this way:

This principle is Custom or Habit. For wherever the repetition of any particular act or operation produces a propensity to renew the same act or operation, without being impelled by any reasoning or process of the understanding, we always say, that this propensity is the effect of *Custom*. By employing that word, we pretend not to have given the ultimate reason of such a propensity. We only point out a principle of human nature, which is universally acknowledged, and which is well known by its effects. Perhaps we can push our enquiries no farther, or pretend to give the cause of this cause; but must rest contented with it as the ultimate principle, which we can assign, of all our conclusions from experience . . . Custom, then, is the great guide of human life. It is that principle alone which renders our experience useful to us, and makes us expect, for the future, a similar train of events with those which have appeared in the past. Without the influence of custom, we should be entirely ignorant of every matter of fact beyond what is immediately present to the memory and senses. We should never know how to adjust means to ends, or to employ our natural powers in the production of any effect. There would be an end at once of all action, as well as of the chief part of speculation.[27]

Hume is not referring to the fact that beliefs, which gather the force of habit, are given credence for no better reason than that others hold them; but to the more fundamental fact that the repetition of experience, by dint of repetition alone, creates habits of thought about cause and effect. What has always happened, we are impelled to think, will continue to happen.

Hume's concern was knowledge and how it is related to experience. Mine is the circumstances in which behavior gains some permanence. But the explanation he gives is as good for my purpose as for his. For if custom "makes us expect, for the future, a similar train of events which have appeared in the past,"[28] such an expectation should exist for behavior as well as for a cause and an effect; I think it does so sufficiently to justify making the "Hume principle" one of the basic principles of sociology and psychology. Hume's repetitions constitute a sort of collective memory, which bears a resemblance to the conceptions of both Halbwachs and Jung. It is usually unconscious, as long as the legitimacy of repeating the past goes without question. But it can become conscious, present to the minds of the actors, whenever there is a challenge. The answer to the question "Quo warranto?" has to be dredged forth from the source of all legitimacy, namely the past. The legitimation comes from the precedents.

These precedents are not necessarily ready-made. They are constantly sought and reworked to fit them for their supporting, and sometimes creative, role in the present. In this the collective memory is very like the individual memory. Compared to habit, and partly because it has to contend with habit, memory can often be a highly active, effortful process. Although some memories are always floating into the mind unbidden, others have to be deliberately and consciously recalled. Whatever their route into consciousness, they need to be organized into patterns so that they make some kind of continuing sense in an everchanging present. This recovery is selective; only the relevant bits of some present which has become the past are recalled to do service currently. In his famous book on remembering, F. C. Bartlett said, "Alike with the individual, and with the group, the past is being continually re-made, reconstructed in the interests of the present."[29] E. H. Carr made the same point about historians:

> We can view the past, and achieve our understanding of the past, only through the eyes of the present. The historian is of his own age, and is bound to it by the conditions of human existence. The very words which he uses—words like democracy, empire, war, revolution—have current connotations from which he cannot divorce them ... The names by which successive French historians have described the Parisian crowds which played so prominent a role in the French revolution—les sans-culottes, le peuple, la canaille, les bras-nus—are all, for those who know the rules of the game, manifestos of a political affiliation and of a particular interpretation.[30]

Remembering is seldom effortless, and it is not at all so when it is called upon to support a whole complex of social recurrences. Institutional cycles tend to perpetuate themselves, like James's flywheels, but they have to be modified almost constantly to keep the wheels flying. This can best be done by trying to remember how the cycles operated in the past and considering how appropriate the old customs or habits still are. Independent thinking about habitual behavior, whether individual or collective, is necessary for stability. It would be an error to imagine that any society, even one as repetitive as the Maya, can maintain itself without constant effort. The legitimacy of cyclical behavior may follow from its ancient origin; but the legitimacy can always be challenged and custom has always to be defended and sometimes remade, just like individual memories. A religious calendar, or any calendar, would be a good deal more fragile if it rested only on habit, powerful as it is. Memory is needed too, to reconstruct the history of the practice with the aid of speech and writing; then, if it still makes sense, memory and reason can reinforce the cyclic with the linear, sometimes in a manner that is decisive, or change it in an even more decisive manner.

In each domain where custom holds sway—such as the law, arts, politics, commerce and industry, welfare, the family, community life, even technology and science—there are learned people who know the history of their own traditions and can restate the old in an authoritative manner, or help to change the old when, in a few particulars or in many, it no longer seems appropriate. People with aspirations to power over others are always reinterpreting old customs to their own advantage, trying to shift them so that others will accept their own authority both as a new fact in their society and as properly time-hallowed. Everyone who innovates builds on what people are already prepared to accept. It is not only in England—though it is exaggeratedly true there, to the point of caricature—that the best way to introduce an innovation is to pretend that it is not one. But the person who succeeds with the new—a Bessemer, a Ford, a Honda, a Franklin or an Eleanor Roosevelt, a Joyce or a Gabo—also alters in some small respect the ongoing customs of the host society with its always conservative partiality.

The Architect of Society

I do not follow Hume in one thing: lumping custom and habit together. Although custom (or tradition, or culture, if those words are preferred) operates through habit, it operates at a higher, more generalized level. If habits aggregate into custom, custom lays down what habits should be and organizes them to be complementary rather than contradictory. Custom is both architect and policeman of society, immobilizing time in a myriad characteristic ways, consigning itself to the collective unconscious so as to allow relatively effortless and cyclical reproduction of past behavior in the present. Always building on the same mimetic tendency, which operates both endogenously and exogenously, custom makes up and perpetuates the structure of a society by making up and perpetuating the personalities of individuals. The constituent habits of the individual are both shared and idiosyncratic.

Custom performs this work by permeating everyone in a society, irrespective of his or her position within it. The language which is the outstanding common property of any society is ordinarily shared by all, whatever their age or sex or role; a common territory usually goes with the language; and there are many attitudes and traits which are in different degrees common to everyone. Customs have the force of shared attitudes and traits, but, more important, each habit and custom can draw upon a general reserve in the mind. People harbor particular affinities for particular customs as well as a generalized disposition to support whatever has been done in the past. What has been is what should be. Habits and customs of all kinds are therefore self-perpetuating, because they generate both negative and positive reinforcements—negative insofar as people feel uncomfortable (sometimes acutely so) when they depart from them, positive insofar as observing them without question produces something near to contentment. So strong is this disposition to custom that, whenever there is a challenge to the customary, people's first reaction is to find out the status quo ante and reaffirm it, if with some tiny modification. The most telling argument in almost any debate about what should happen is that it is in accord with the practice of the Constitution or any constitution, of the Gospel or of a gospel. The presumption is in favor of the past.

The binding force of custom is continually recreated by cyclical behavior. Members of a group interacting together again and again in the same fashion generate not only solidarity with each other but also solidarity on behalf of the recurrent behavior itself. The sentiments created when one person acts upon another can be stronger and more readily articulated than the sentiments associated with a purely individual habit, though that can be tyrannous enough. People form attachments not only to each other but to a house and a district; the layout of each room and the furniture in it and the view from the windows, the turning of each street or the trembling violet silhouette of particular trees against the evening sky become incorporated into a body of habitual feelings, and people feel grief and nostalgia if they have to depart. The attachment is reinforced by the attachment of other people to the same location. The homesickness is infectious. By and large, the more frequently people interact together, in a district, a workplace, or any other setting, the more positively as well as negatively charged solidarity they feel for each other. When a whole nation is challenged by another, we know only too well what follies people are eager to commit, how easily they can stoke in each other a fierce heat of loyalty to their own customs and the symbols that stand for them and of hatred for those whose customs are made to appear so outrageously different. Their very identity seems more at stake in the collective than in their individual habits. On this score people are notoriously altruistic: more willing to die, and kill, for the collective than they are for themselves.

Custom does not work only in this blanket manner, right across society, but focuses with some precision on individual people. Custom can be highly differentiated. If it were not, if all people were as alike as mimetic clones, there would be no variety, no stimulus, and no recognizably different societies. As it is, custom hangs itself on the pegs provided by biology, especially age and sex, and weaves other patterns which have nothing to do with biology. Customs have great variety, forming a "pool of habits" for a society like the gene pool in a genetically isolated unit of population, inducing in people not the same habits so much as complementary ones. Reciprocal habits are more stable than purely individual habits. They fit with each other by being different, and it is the differences which create lively and lasting relationships. The structure of a society, rather like the structure of a chemical compound, is in good part maintained by the bonding of opposites, or at any rate of complementarities.

Here then is yet another analogy between society and nature. Unlike electrical charges attract each other; genes are made up of strands, each of which is the other's predetermined complement; custom mimics genetics by creating reciprocal habits and habit-driven conflicts, or building them on the differences reproduced by nature. Such habits are the roles which have been the stock-in-trade of sociology. Men and women are not allocated quite the same roles; nor are young and old, employers and employees, doctors and patients, priests and congregations, teachers and pupils. Without many habits which are the same—above all, the shared grammars, syntax, and vocabularies which make up the languages with which we strive to communicate—there could be no general social order combining all the roles. Without many habits which are unlike but complementary, however, there would be no differentiation in society within an overall unity. If custom ruled out differentiation it would deserve to be styled "the lump of custom," as it often has—but I think it does not so deserve. Custom has delicate work to do, writing so prolifically the parts which people perform with such wonderful versatility, moving in the course of a day, an hour, or even a minute from one role to another—as though in their mental wardrobes hung countless clothes which all of them, only too anxious to be actors, are waiting eagerly to put on, or as though on their music stands lay countless scores which they are waiting eagerly to play yet again. Although many of the parts are decreed, the way they are acted or played is not. People are suggestible, but a lot of the detail can be left to their own invention. They neither have to repeat all the lines or notes nor have to want to. The play or the symphony comes alive, and the actors become playwrights, the players composers, through the energetic reinterpretations which operate in the same active manner as memory.

The natural asymmetry of sex differences is more marked than any other. In every society custom has draped itself, often ponderously, around this particular biological difference, with the habits assigned to men and women often standing in sharp contrast to each other. For the sake of illustration I am going to pick another set of biological differences more immediately relevant to the main subject of this book—that is, the changes related to another biological clock, the life cycle—in order to argue that it is custom which strings together one life onto another in order to create the continuity of society. The life cycle is the series of changes which each person goes through between birth and death, and which becomes cyclical by being repeated in other people. Without such cycles from one generation to another, there would be no persisting society.

The Marching Column

Age differences are pressed into service to bestow one kind of order on all societies. A man and a woman of the same age who

had lost their memories and were isolated on a desert island would
have no notion of the life cycle. They would imagine they would
live forever. The changes in their bodies as they grew older could
be continuously astonishing, and death even more so. They would
not fear what they did not expect, although as they neared death
they might be filled with even more foreboding than if they had
known what was going to happen. But surround them with their
parents and children and other people of different ages on the
island, as they are in every society, and the couple would at once
realize what they had been and what they would become. They
would be surrounded by their tutors, young and old.

I am now getting so close to the analogy of genetic reproduction
that it may seem as though it is already ceasing to be an analogy
at all. But that is not quite so. If biology explains how genes are
perpetuated by being transmitted sexually, sociology should be
able to explain how the social habits which make up societies are
transmitted socially. I think I can best do so by means of the
image of society as a long column on the march. The marchers,
with their different habits, are drawn up and moving forward in
age order.[31]

All the people in the column, whatever their age, are incarcer-
ated in the same continuing present, from which there is no escape
(at least for humanity of the existing design). But if we have direct
and immediate experience only of the present, this present is
layered with the past and the future. People can cast their minds
backward as far as their eyes can see to the babes in arms, and
forward as far as their eyes can see in the opposite direction, to
the older people who are waiting to fall out for good. If the
column were quite stationary, with every old man's cane and every
young child's hoop motionless, as if waiting for the order "quick
march" to bring it to life, the metaphor would suggest only that
a society is made up of people of different ages, whereas I want
to convey the idea of motion in the present, of the ceaseless cyclical
changes which keep things the same. People recognize the motion
by identifying with others. When people look back at the children,
they can see something of themselves when they were younger,
before they changed into what they are now; when they look

forward they can see something of what they will be. They can identify with the others because they are interchangeable with them. In their imaginations they can always change ranks, especially if their gaze is toward the rear of the column. People identify more easily with those of an age they once were, and which they can remember, than with people of an age they have not yet reached. This is one reason, no doubt, why affection descends more readily than it ascends, finding its object in people's own children and even in the children they meet who are so like (and yet so unlike) their parents. Proust describes meeting Mlle. de Saint-Loup:

> She had cleanly-shaped, prominent and penetrating eyes. I noticed that the line of her nose was on the same pattern as her mother's and grandmother's, the base being perfectly straight, and though adorable, was a trifle too long. That peculiar feature would have enabled one to recognise it amongst thousands and I admired Nature for having, like a powerful and original sculptor, effected that decisive stroke of the chisel at exactly the right point as it had in the mother and grandmother. That charming nose, protruding rather like a beak, had the Saint-Loup, not the Swann curve. The soul of the Guermantes' had vanished but the charming head with the piercing eyes of a bird on the wing was poised upon her shoulders and threw me, who had known her father, into a dream. She was so beautiful, so promising. Gaily smiling, she was made out of all the years I had lost; she symbolised my youth.[32]

It is not so congenial to leap forward to the older marchers in whose steps we are following; although fortunately we can never catch up with them we have to reflect that one day we will replace them just as one day we will be replaced. The firm expectation that in the ordinary course of events we will soon be in their place can hardly fail to arouse sympathy for them. If age is a problem it is the same problem for all of us, even if we become steadily more aware of it as time passes.

Because we have no receptor for time, and therefore no good means of "visualizing" it, this is a spatial metaphor, but at least by likening time to a column continuously on the march the metaphor underlines the point that a society is a phenomenon that is crucially temporal. It is also apt because it again contains the two kinds of temporal motion which reinforce each other in society, the linear and the cyclical.

Linearity is shown by the column's not endlessly circling around, forever passing the same reviewing stand, but moving straight ahead. The column moves in three different tempos: slow-march, medium-march, and quick-march. In keeping with the slow tempo, some habits of the marchers are changing very slowly or hardly at all. For this purpose, the line of march is considered as stretching farther than the eye can see in both directions. There are then no cutoff points for the column as there are for the individuals who compose it. Individual marchers are born and die, but the column is immortal, made up not just of living people but of those who are dead and those who are not yet born. The dead taught the older people what they are now teaching the young, and they in turn were taught by those even longer dead. Many of the tools the marchers carry date well back into the past, as does the language in which they mutter to each other. Every word we use has been through a million minds of people now dead before it reaches ours, just as water from an urban river goes through many different stomachs as it is time and again taken out of the river for drinking and put back into the river as effluent before being recycled. It is not necessary for new recruits to invent a whole new vocabulary, nor reinvent the wheel or the means of controlling fire.

New customs or variations on old customs are, of course, continuously being invented by exceptional or lucky individuals, who do so quite deliberately or as the result of random error in the copying of old customs. Some of the crucial changes may be abrupt or revolutionary: people may push a series of innovations so far that a kind of "speciation" occurs, marked not by the erection of biological barriers to interbreeding between two groups that were once one but by the erection of barriers to mutual solidarity and comprehensibility. This can happen when a new language is created or occasionally by dividing older against younger members of a society.[33] But much the greater part of what is learned by the new recruits is handed down from the distant as well as the immediate past, and the greater part of what those not yet born will learn is already known. Because custom is cumulative, the general attitudes and the general knowledge as well as the objects which embody them are of long standing, and will so remain.[34] Whispers pervade the air above the column: some of the whispers certainly come from the dead, and, without unprecedented changes in human practice or unless a catastrophe wipes out all the living marchers in the column, some of them will reach the ears of people 2,000 or 20,000 years hence.

I called the second tempo medium. Although most of the behavior displayed by the marchers is the same as that of others now dead, they are aware that some of it has changed since they were born. Fertility, for instance, is always changing, and it has done so most markedly within current living memory.[35] There are no longer so many babies at the rear of the column, nor so many of them dying as infants. The reduction in fertility has in turn encouraged more women to enter the labor force and to assert rights denied to them while they were still bound to childbearing. There have been many other changes almost as significant, such as improvements in health care and nutrition which have increased the relative number of old people, and especially old women, in the population or (as it might be put) enabled them to stay "young" longer; the increasing fragility of marriage which has left more parents to cope single-handedly with their children; and the lengthening of childhood as the years spent in formal education have been extended. These demographic changes have been contemporary with hundreds of others in attitudes, in the standard of living, and in the style of life. In these conditions of constant change it makes some sense to talk of the "fluid life cycle."[36]

The third and fastest tempo applies to the individual alone. The rate of change is relatively enormous in the early years: the child develops new capacities continuously, learns new skills, and acquires both the customs which have changed at the slowest rate and those which are of more recent origin. But even when the habits acquired so rapidly have settled into relatively fixed personalities, individual people can and do change relatively faster than the whole society of which they are part, if only because they age (and change with age) while society, though it undergoes secular change, does not age so long as the age composition of its members remains the same.[37]

Social Metabolism

The marching column metaphor is equally, or perhaps more, apt for the cyclical dimension. If people lived forever this would not be so; nothing would be manifested except linearity, and the general term "life cycle" would be discarded in favor of "life line." But because people die, they have to be continually replaced. A leading anthropologist, A. R. Radcliffe-Brown, has described this process using the same analogy I do, except that he has displaced my genes with his molecules: "A nation, a tribe, a clan, a body such as the French Academy, or such as the Roman Church, can continue in existence as an arrangement of persons though the personnel, the units of which each is composed, changes from time to time. There is continuity of the structure, just as a human body, of which the components are molecules, preserves a continuity of structure though the actual molecules of which the body consists are continually changing."[38] This seems to be grippingly true of the sphere which the Radcliffe-Browns inhabit: not only do Radcliffe-Browns seem to last unchanged longer than any of the molecules which make them up, but institutions like university departments of anthropology last unchanged much longer than any of the Radcliffe-Browns who compose them.

The social metabolism requires more than just producing roughly enough children to replace the older people who die. It would not be enough to have just as many cradles as graves lining the route: babies cannot go straight to university or into Mr. Murgatroyd's machine shop without seeming out of place. If the column is to retain its characteristic age order, succession has to be by age group, children succeeded by other young children, young adults by other young adults, the middle-aged by others like them, even the old by the old. In every age group dropouts have to be offset by drop-ins. Biology is responsible for people being of different ages and determines the potential for growth and decay, but it is not responsible for people of a particular age doing more or less the same things as other people of that age before them. The succession of age-specific habits is a sociological matter. It is a matter of custom.

The process cannot be explained in the military terms of an ordinary army, and if the metaphor suggests it can it is misleading. There is no commanding officer, no general standing on a rostrum shouting his orders up and down the column. Everyone plods along in the column, a commanding officer as well as a subordinate. Some have more power than others, but in essence everyone is keeping everyone else in place, by precept, by sanction, and above all by example; as I said earlier the different homeorhythmic processes keep each other in order. Common customs prevail only because they are carried by the many members of a society almost as surely as common genes are carried by all the members of a species. The column is therefore composed of civilians rather than soldiers, or if it is military at all it is made up of soldiers like those envisaged by the Duke of Wellington, whose habits are so much "tenth nature" they no longer need to be told what to do.

When they are still quite young, children are enjoined by their parents to "act your age" or shamed for being "babyish" when they are no longer supposed to be babies. The same injunction not to be a "baby" or a "child" or "juvenile" is leveled at people of any age, with a unique precision in industrial societies provided by the registration of births and the resulting control over what people can do at certain ages. In societies without written records most people cannot know just when they were born. They do not mind in the least if they cannot reply to the question "How old are you?" We almost volunteer the answer on hundreds of written forms and in hundreds of conversations. The people who do mind the lack of birth records in paperless societies are the officials who have the thankless task of trying to conduct a census.[39] Without registration of births, such societies also lack the informal ritual about the celebration of birthdays, which does what Aristotle said legislators aim to do with legislation, that is convert it into custom.[40] Every parent who organizes a birthday party is an unwitting agent of the state, even if its power is masked by tying every birthday to the annual "life cycle" of the sun. The state and the sun make a powerful alliance. Without the bureaucracy of modern societies the only way people can calculate their ages is to link their birth to some contemporary dramatic event in the collective life, such as an earthquake, a flood, or the death of a chief or a king. This was how the nurse in *Romeo and Juliet* calculated Juliet's age—"'Tis since the earthquake now eleven years," by which time she had already "run and waddled all about."[41]

In our column a whole apparatus of control, now encrusted with tradition, has been hung onto the bureaucratic reckoning of age. Everyone is conscripted into a linear system of measurement in order to make the cyclical replacement more precise, with no conscientious objection tolerated. The registration by age becomes one of the most significant facts about people: the book determines when they are supposed to enter school, leave school, marry, drink, vote, smoke, get called up for military duty in a real column, draw a retirement pension, and a great deal else in between the registered birth and the registered death. All the marchers carry on their backs a tag with this most important number on it. Every year you add one to the number. It is as if you wear a watch with a limited life, which starts at zero and is then set forward once a year until it reaches its final number and stops. Society compels people to display the watch on demand as if it were a pass they always had to carry.

Partly as a result of the fine distinctions the arrangement allows, other reckonings proliferate. People become highly skilled at "averaging," acquiring a finely drawn portrait in their minds of the way a person of sixteen, forty, or sixty-six, looks or behaves, so that they can say without a moment's hesitation as soon as they know a person's registered age, "She does not look her age," or "He looks older than his years," or "She acts more like fifty-six than sixty-six." Each individual's judgment is honed by frequent checks with nearby marchers, so that all can with facility paint this generalized portrait that a person of fifty, or any other age, should resemble. In certain circumstances it can invite as much ridicule to have a young head on old shoulders as to have an old head on young shoulders, to be lamb dressed like mutton as to be mutton dressed like lamb.

The bureaucratization of aging produces many abrupt qualitative leaps from quantitative changes, the kind of which dialectical materialism is so fond. In Britain, after five birthdays have passed, children are whisked away from home and forced into an enormous building full of strangers; after twelve more, pulled out of the building and into the unemployment queue or a job; and after so many more, put out of a job for good. The criterion is not developmental age in biological or any other terms but simply chronological age. This schedule is obeyed far more because it is custom than because the law enjoins it, although law and custom reinforce each other.

Parents start the crucial training long before children reach the age of five by performing another sort of averaging. Children race through eons of development in a year or so. Babies do not walk out of the womb as a crocodile walks out of the egg. They have to learn to move their limbs and to crawl before they can triumphantly stand on their hind legs, and babble and coo before they as triumphantly learn to talk. No later achievements will ever be so spectacular as these. But for the parents who provide the conducive environment the wonder can be marred if their child does not walk or talk or do dozens of other things as early as other children. The delight of seeing a child stagger independently across the room, or cave, with no arm to lean and learn on, may be lessened if other children have done their staggering even younger. Parents may fear their own child is "backward."

The pressure is certainly not removed when children go to school. Transmission of culture is still mainly vertical, from the older to the younger.[42] Teachers join (or replace) parents in order to pass on the fears, knowledge, and practices appropriate to the age of the pupils. But because schools are usually organized in age groups, horizontal transmission begins to take over. The practice of "family grouping" when children of different ages are in a class together is rare. It is more usual to lump pupils together according to the date of their birthdays. This produces curious injustices, with summer-born children being put at a permanent disadvantage in Britain because they enter their infant school in September, whereas those born between September and December enter in January and have two more terms at the school before they go up to the next school which they enter at the same time as the summer-born children, over whom they have a head start.[43]

Unjust or not, the year group becomes for many purposes the group with whom people march. One nineteen-year-old carpenter in Bethnal Green in East London described his peer group this way:

> There was a group of four or five of us who always used to go about together. One of them I'd known since before I went to school. We knew each other's mothers very well—I called his Mum "Auntie" and he called my Mum "Auntie", sort of thing. We used to play together when we were little kids. Then we went to school together and we met other kids at school. It was always just boys— when we were about seven we used to play rounders with the girls in the street, but that's all. There was never any leaders—we decided things together. We used to play football together in the street and go over the park. Then we started going to the pictures together or swimming or playing football with other kids, that sort of thing. When we got to about 12 or 13 we used to lark about—we used to thieve a bit, now and again, but nothing serious. We went out together most nights and we all went to the same youth club when we were about 15.[44]

Each such horizontal group is also held in place by more or less vertical transmission from other year groups, both older and younger. The children at the top of the school (the word meaning more than just top in terms of age) can seem almost "grown-up" to those who have just entered at the bottom. They can have the same influence as older siblings, especially if they are of the same sex. The older pupils who play sports for the school are models for the younger ones who, if they play at all, play for their class, and those in the older class are always ready to rebuke the younger ones for any pretensions above their station. The older are all the more effective in this role because their example is not to be followed immediately but only after a delay, when the younger reach the right age for replacing the older ones in a certain class or gang, in having a certain amount of pocket money to spend, in watching their own special TV program, in appreciating the batmen or spidermen, spaceships or robots in their own age-specific fairy stories and toys, in wearing the clothes which are proper for their age, and in cultivating the small distinctions in speech and manner which bulk so large to those learning the age-maze which everyone has to wind through on the route to adult-hood.

As they grow, the age group with which children feel an affinity gets steadily larger. At the age of five or six it may be only a few local friends, but by adolescence a good part of the world may be symbolically represented in their mental living space. The fashions they follow are usually differentiated between one age group and another. People on the margin between youth and adult society seldom display the same fashions as the younger but usually find it difficult to drop their partiality for what was in fashion when they were that age. If they looked exactly like members of a younger generation they would run the dreadful risk of losing their age tags and so being altogether lost.

The cult of youth, however, crosses all boundaries and draws into its sphere many who are no longer young, except in aspiration. At odds with my genetic analogy, the standard vertical transmission is sometimes reversed: older people copy younger. Karl Mannheim remarked: "The extent to which the problems of younger generations are reflected back upon the older one becomes greater in the measure that the dynamism of society increases. Static conditions make for attitudes of piety—the younger generation tends to adapt itself to the older, even to the point of making itself appear older. With the strengthening of the social dynamic, however, the older generation becomes increasingly receptive to influences from the younger."[45] This applies not only to fashions in clothes but to fashions in sex and fashions in ideas, and the more creative youth is, the more agitated the attempt to jump ahead instead of keeping soberly in rank. In taking a fresh view of custom, younger people often have support from their elders. The standard sort of vertical transmission is put into reverse: parents are disappointed if their children have no ideas different from their own, holding them at fault for being old before their time. Custom once again plays on the differences between people of different ages, translating differences in biological age into differences in social age, which provides the leverage to transmit a crucial kind of inheritance in both directions.

Back to the Lockstep

Age-specific customs are generally most strenuously binding in the first period of dependency, when they are strongly reinforced by the state. Industrial societies have not only made of modern childhood a long and distinct period of life but have given the state the prime responsibility for regulating it. After childhood, when specifications by age become less precise, it may be more appropriate to talk about stages than ages. It then matters more whether people have entered the stage of getting married and having children of their own than how old they are.

Such a holiday from age typing does not last long. The custom of marriage is still followed by most people in the end. Even now, after all the relaxation there has been in the customs of family life, some young women and men worry that if they do not marry soon and have children they may miss the chance, and have no option but to remain single and childless. Women in their late thirties and forties who want children but have not yet had them find biology once again bearing down on them. If they do marry and have children, their children tie them into the same lockstep of childhood that engaged them when they were young.

The occupational structure has many age-differentiated customs of its own. People who enter an occupation that they hope to stay in soon begin to copy the habits of those older than themselves. James wrote in 1890 about the professional manner: "Already at the age of twenty-five you see the professional mannerism settling down on the young commercial traveller, on the young doctor, on the young minister, on the young counsellor-at-law. You see the little lines of cleavage running through the character, the tricks of thought, the prejudices, the ways of the 'shop,' in a word, from which the man can by-and-by no more escape than his coat-sleeve can suddenly fall into a new set of folds."[46] James was writing about Boston toward the end of the last century; but he need not have been. It has been said that while there's death there's hope, but it never changes anything as long as the dead man's shoes are filled, as they often are, by a near replica of himself, and the near replica joins so enthusiastically in grooming himself for the identical part.

In professional or managerial jobs the adopted mannerisms serve the purpose of helping people to advance their careers, another temporal notion which has become of the utmost importance to those who believe they have them. A career, according to Wilensky, is a "succession of related jobs, arranged in a hierarchy of prestige, through which persons move in an ordered, predictable sequence."[47] There is a recognized maximum number of years anyone should stay on a particular rung on the ladder. If an aspirer stays less than that, the chances are good; if more, bad. In a metronomic society it can be measured. Mr. Farrow, a manager I interviewed for a survey I made with Peter Willmott, said, "In my firm they are good about that. If you've got it in you, you can climb right to the top. In five years' time, if I keep going, I should be Deputy Sales Manager for the whole of the South-Eastern Region. I might end up as General Sales Manager, and even get on the Board—but frankly I don't think that's likely. If I was going to get there, I ought to be higher than I am now, at 35."[48] Ambitious executives try to read the oracle of advancement until their "career menopause," when they have to accept that they have attained the highest point they will ever reach. At each stage in their careers people calculate their progress in relation to all the other striving people around them, rather as they simultaneously calculate their children's progress in their vital educational career. A career is a special kind of custom which both children and their parents seem only too happy to cling to, as long as some hope of linear progression remains.

After a career is over, a new stage is reached when people retire from paid work. The notion of retirement itself is a novelty: a general phenomenon only in the twentieth century. But as habit begins the first time, so does custom. New customs are still developing about the age for people to retire into what has been called the "roleless role." In Britain it was accepted until recently that retirement should be fixed by the state, and that age established a norm for employees and employers. People not only could afford to retire then, if the state pension was adequate, but they *should* retire at that age to make room for the younger people pushing their way forward in the column behind them. The practice was hallowed by the many people who had previously retired at that age. Now the custom has begun to change, as increasing numbers of people retire earlier than the age that was customary, and as unemployment for older people merges with early retirement.

People do not have a free hand to decide to retire whenever they want, however. They take their cues from others in the column, not just about the age at which they will be entitled to financial support when their wage ceases but the age at which it will be "proper" in a broader sense to stop work, with no shadow of blame falling on them for not earning their own keep. About such a matter there has to be reciprocity of the kind that custom embodies. People cannot decide on their own that they count as "dependents"—young, old, or ill—who are entitled to support; the givers of support have to agree too. New variations on the familiar custom of retirement have to be worked out by both parties before the retired can make a new life, compounded of old habits which can be adapted to the new stage of life and new habits which can take the place of the old.

All societies have customs about age. The features most marked in our society are the precision and extent of measurement. These are apt to provoke anxiety. As Maurice Halbwachs said in an excellent sociological study, "What is probably more painful is that I feel perpetually constrained to consider life, and the events filling it, in terms of measurement. I anxiously reflect on my age, expressed as the number of years lived and remaining . . . as if the years ahead of me shrank in proportion as the elapsed time of my life increased."[49] The lack of escape from biological aging encourages people to believe there is also no escape from the customs associated with social aging. They are like the seasonal rituals which are compelling because they are associated with nature. Custom is twice custom if it can enlist nature on its side, and can be twice wounding if it compels people to slice up the continuous process of aging into a series of measured intervals.

Authority

I have drawn attention more to the force of custom than to the authority possessed by parents, teachers, managers, and the state.

But without invoking authority, it is not possible to unravel the Murgatroyd puzzle. Why do the machine operators put up with the unnatural cycles which are their lot? It cannot be that Mr. Murgatroyd is not Puss in Boots but Apollo in boots, striding fiercely between his machines rather than riding the heavens in his celestial car. If he has any charisma it is not (like that of rulers before and after the Roi Soleil) borrowed from the sun; no one would be more surprised than Mr. M. by any such double comparison. Nor is it just custom that keeps the backs of his men bent, though they would not bend them without it. It is authority. Because Mr. M. can deprive them of their livelihood, when he shouts at them they can only curse back under their breath.

A further edge can be given to the puzzle of authority by an account of a laboratory experiment.

In a psychological experiment conducted at Stanford University students who took part volunteered to enter a mock prison in the basement of the Psychology Building. Chosen at random, eleven were to be guards and ten prisoners, and to follow something like the routines of ordinary prisons. The guards were not "trained" but it was found that very quickly they knew what to do: they only had to repeat the behaviour they had seen many times in movies and on TV and read about in books. Brutality came easily to them. The plan was for a two-week simulation. But a real prison was created so rapidly that it had to be aborted after only six days, and even before that four of the prisoners had to be released because of extreme emotional depression or acute anxiety attacks or because of a psychosomatic rash over their entire bodies.

It was not long before the guards began to demonstrate their inventiveness in the application of arbitrary power. They made the prisoners obey petty, meaningless and often inconsistent rules, and forced them to engage in tedious, useless work, such as moving cartons back and forth between closets and picking thorns out of their blankets for hours on end. Not only did the prisoners have to sing songs or laugh or refrain from smiling on command, but they were also encouraged to curse and vilify each other publicly during some of the counts. They sounded off their numbers endlessly and were repeatedly made to do pushups, on occasion with a guard stepping on them or a prisoner sitting on them.

The prisoners played the part as convincingly as the guards, and after two days it was no longer a part. It was them. They had become brutalised, not by a lifetime but by a few days in which they conditioned themselves according to what they believed the experimenter wanted them to do.[50]

Such behavior, examples of which could be multiplied, cannot be attributed to custom alone. Custom is no doubt, as Bacon said, the magistrate of human affairs, and a magistrate who is obeyed so readily because people are not aware they are doing so. But the magistrate's writ does not reach everywhere. Not all the rules that people follow are laid down by custom. Some are imposed by authority.

Customs do not always arise from freely chosen imitation; people are often told flatly whom and what to imitate. Parents tell their children who the ideal models are, namely themselves, and what to do, as well as rely on their learning what to do through mimicry; so do teachers and many other adults who crowd their own feet into the parents' shoes. As a result many people become so accustomed to being told what to do that they feel quite lost if left to their own devices.[51] For them any person with authority is invested with the authority of custom; such a person is still more awesome if the custom has the aura of tradition to it. The customary power of authority is built up by being repeatedly exercised and handed on from one holder to another.

If authority bolsters custom, custom bolsters authority. That additional flywheel is always turning. Recurrences may have to be initiated by an order but as they are repeated they may become second or tenth nature, and their necessity accepted almost as though they belonged to the natural order: as though Puss is Apollo. People in authority strive for this effect by establishing recurrences and intentionally creating other people's habits. Just as it is better to have a dog trained in obedience than one which has to be kept on the lead, every manager knows that his power will be more complete if his subordinates can be persuaded they are acting under their own orders rather than under his. If managers can exert that degree of influence over the minds of their subordinates, they can rule in the ideal state of consent.

This power can even continue to exist long after the setting in which it was relevant has been left behind. One of the workers from the car factory who retired early had endured a lifetime of work so gruelling and repetitive one would think he would be heartily relieved to put it all behind him. Far from it. He pined to be back under the lash of Mr. M.'s tongue almost as soon as he had cast off his working clothes forever—or (one might say) as soon as he found he could never cast them off. There are unemployed people, like those studied by Paul Lazarsfeld and Marie Jahoda, who are miserable not only because they are without a wage, but also because they miss the firm time structure they had when they were in the grip of authority. After a period, even the new routines they invent are no longer comforting. When prisoners are released they may long to go back to the little daily routines which had meant so much to them in their cells. James put much of the subordination in society down to habit: "It alone prevents the hardest and most repulsive walks of life from being deserted by those brought up to tread therein. It keeps the fisherman and the deck-hand at sea through the winter; it holds the miner in his darkness, and nails the countryman to his log-cabin and his lonely farm through all the months of snow."[52]

But authority is also at work, often relying on a biological cycle, but in an oppressive way: neither shall he obey, neither shall he eat. Authority is invoked only when custom is not adequate on its own, and although authority can be so much blended with custom that it is completely disguised, there is also a continuous transformation in the other direction, of custom into authority. Authority is a kind of catch-all or fallback for custom which is drawn on when (as increasingly happens in an ever more complex society) custom no longer suffices on its own.

More Than an Analogy?

The main idea of the analogy from molecular biology has been that as the genes guide the reproduction of the organism, so a disposition to recurrent behavior guides the reproduction of habits. This assumes that habits can be regarded as a single phenomenon despite the existence of so many different sorts of them.[53] The 1930 *Encyclopedia of Social Sciences* differentiated among motor habits, cognitive habits, emotional habits, and moral habits; if the word had not been dropped from future editions in response to fashion, the categories could be further multiplied. These types of habit nonetheless have a common character: in all cases some aspect of past behavior, individual or social, is replicated in the present. The great society is habit writ large. Emerson could have been writing in an age of molecular biology when he said, "The life of man is a self-evolving circle, which, from a ring imperceptibly small, rushes on all sides outwards to new and larger circles, and that without end."[54]

In biology a distinction is made between the phenotype, the behavior, and the genotype, its genetic constitution. In the analogy a habit is like the phenotype and the disposition to habit or custom like the genotype. But perhaps habit is more than *like* a phenotype. Could the process be not analogous but the self-same, with genes giving rise to the disposition? Before trying to answer, I should be a little more precise about the nature of the disposition. I do not mean by it just the tendency of people to replicate themselves by persisting in a habit once it has been adopted, but also the tendency people have to imitate each other. If they do not know how to behave, they look for models in other people, which makes custom more like sexual reproduction than like purely individual metabolism. Of course more than two people are involved; a custom can have many progenitors. Whereas genes are replicated from generation to generation by the mixing of those from the two parents, custom can be replicated through a nationwide or even worldwide "polygamy" of hundreds or thousands or millions of other people contributing to a "custom pool" that is more like an ocean.

I call the reproduction of custom a disposition or a tendency rather than a rule because it does not always have to be followed. Although the pressure to replicate may always be present, it can be repressed in any particular case. Reason acts as the repressor when a habit proposed by the social environment, or even the natural environment, is judged inappropriate. Indeed, problematic as it is in practice, one of the main functions of reason is to repress the adoption and to suppress the continuation of unsuitable habits, as well as to support suitable habits.[55] If genes are involved, they will have to be switched on by the recurrences: but they need not be.

The disposition to cycles is evidently a common property of human beings. People are constantly ready to replicate behavior, to hunt for eggs at Easter or place candles on Christmas trees for no better reason than (or for the good Humean and human "reason") that "we've always done it that way." A name can be given to what is replicated that suggests the similarity with genes: the biologist Richard Dawkins has proposed that the "unit of cultural transmission" be called a "meme." Examples would be tunes, ideas, catch phrases, clothing fashions, ways of domesticating animals or building arches. "Just as genes propagate themselves in the gene pool by leaping from body to body via sperms or eggs so memes propagate themselves in the meme pool by leaping from brain to brain via a process which, in the broad sense, can be called imitation."[56] This seems to me to go too far by suggesting that there is a particular unit in the sphere of custom and culture which is the same as the unit in the genes, whereas it is surely more sensible to talk about a general propensity. Human behavior is not quite like that of Konrad Lorenz's geese, who became imprinted on whatever they saw with the right characteristics at a critical stage in their development. With human beings, whether or not it can be pinned down as a "unit," anything whatever can be imitated at any stage and at any age, although early impressions are the most indelible. This may be because maturation is so much more drawn out than in other animals and in a sense is never completed. The proper analogy is with the process of reproduction, not with what is reproduced. But in describing the overall process as one of imitation I am entirely happy to imprint myself on Dawkins.

The disposition is useful. It gives people (and all other creatures which have a capacity to learn from their experience) the power to encapsulate the past. They can use it to build on other people's knowledge and on their own. But it does not necessarily follow that there is a genetic substratum to it. People could repeat themselves because they have learned the hard way, through experience, the advantage not only of particular habits but of a generalized disposition to form them. Whatever their genetic constitution, this lesson is reasonably simple to learn. Even a stream "learns" to follow the course of least resistance; a lock works more smoothly the more often it has been turned; a car goes better after it has been driven for a while; new jeans become more comfortable after they have been worn in. Streams, locks, cars, and jeans do not contain genes for this purpose, so why should people? We could continually fall back on repeating ourselves because it requires the least effort; and that is no doubt a factor. All living systems need to minimize the consumption of energy. But this is done in different ways, and I do not think that with human beings effort-conservation is all there is to it.

If there is a general disposition to cyclicity which derives solely from experience since birth it could coexist with other more specific dispositions, triggered by genes which enable children to suck, smile, walk, and speak, and also to make use of their biological clocks. Noam Chomsky has made a case for his "innateness hypothesis" about language on behalf of a Universal Grammar: "Linguistic theory, the theory of UG . . . is an innate property of the human mind. In principle, we should be able to account for it in terms of human biology."[57] Once habits (or instincts) have been used to lodge particular capacities in growing infants, they could be replaced by a general disposition. The operation of specialized habits could leave behind a generalized capacity to learn in this same manner.

The process could also have happened the other way round. A generalized propensity to copy could be used for an endless variety of specific purposes. The skills of walking and speaking could be acquired with such astonishing rapidity by young children because they can call on this generalized propensity at an age when they are highly malleable. A disposition formed this way would still be incident to an "innate property of the human mind," but to a general rather than a specific property. This is possible but not likely. In the case of language children acquire a skill beyond anything they could have learned by copying others at that age, even if they have to copy others in order to acquire it at all.

Although I do not want to exclude specific propensities, I think it is more likely that there is a general genetic predisposition which can perhaps explain Hume's "principle of human nature" and allow his enquiries now to be pushed further. There are four main arguments.

1. I mentioned the discovery of the per genes in Chapter 2. These genes do not require a completely fixed response. For example, per genes maintain free-running rhythms even while being entrained to a 24-hour periodicity by the environment. They have that small degree of freedom. The genes I am presupposing would be like per genes; or they might be per genes which in human beings have mutated so that they are capable of being "switched on" not just by the sun but by any regularities in the environment. They could be entrained or disengaged by any regularities, although they need not be. Further study of per genes may show whether they are responsible for habits. Perhaps per genes will prove to enable an organism to harmonize with any environmental cycle and also dispose it to seek out cycles.

2. If I was right earlier on when I suggested that habits can be considered part of memory, the likelihood of the genetic explanation is surely improved. For the capacity for memory (as distinct from what goes into it) could hardly be acquired purely as the result of experience. It is present before there has been time to gain much experience; babies can recognize a face and a voice not long after birth. Recognition rests on memory and becomes habit. The accessible programs stored in memory can evidently be retrieved in two modes—act or enter consciousness—by a sort of printing. When the mode is act, they may not be printed. When the mode is print, they may not be acted upon. It would be strange if the two modes of remembering did not have the same origin. If we accept a genetic base for memory, it is not easy to reject it for habit. Babies are also fond of habitual, reiterative play, which is the means by which they learn.[58]

3. It is clearly more economic to go on doing the same thing (varied where necessary) than to draw back in every situation and puzzle over what to do next. But it is not so clear why, before people have acquired a habit at all, they should be so ready to adopt those of others. Other people's habits are acquired as the result of their experience, not one's own. Why should one be so

ready to accept on trust what they have learned? I can point up the question by recalling a small three-year-old boy whom I watched on his first day at a nursery school in London. At first he was lost, looking around anxiously for his mother or the teacher and wandering aimlessly and tearfully around the classroom. But after a bit he returned to one corner where four other children were playing with water and sand in a fiberglass pool. Eventually he joined them and started splashing the water about as vigorously as they did and trying to make sand pies as solid as theirs; and forgot himself. He was soon chortling with as much gusto as any of them. He was at home. He had learned to play by doing exactly what was being suggested by the others. His suggestibility is shared in some degree by the rest of the human race. I find it difficult to resist the notion that such a basic suggestibility, which is so essential to our species, must come from the book of nature as well as from the book of nurture.

4. The main reason for proposing a genetic explanation is that habits would have given—in my view have given—a vital selective advantage to the species. In competition with creatures with much more fixed programs, human beings depend upon being able to learn from their experience, and to learn fast. This they can do so well because they do not have to relearn what they have already learned. Habit is a means of making learning stick. What they have already learned can be consigned to a store from which it can be readily retrieved, more reliably and with less effort than if it all had to be recalled from memory. They do not have to restock the store constantly but can concentrate their effort upon learning the extra bit which is added to it at the margin. Habits therefore have a function in relation to new experience as well as to old. I have made this point before but want to stress it again.

I agree with James that habits and customs are "the most precious conservative agent" of our individual and social personalities. But more than that, they are the means of dealing with not only the predictable but also the unpredictable. Any recurrent event calling for a response can generate a disposition, and the degree of its entrenchment ordinarily depends upon how often the event has recurred and the standard response has been appropriate. There seems to be a kind of unconscious, rough-and-ready probability assessment, based on the rule that the greater the frequency of recurrences, the more probable that the recurrences will continue and the less the risk in responding to them in an habitual manner. The greater the frequency, the more deeply entrenched the habits become. They may go so deep, especially if the earliest instances were in childhood, that they cannot be set aside at all, in which case an effort may be made to adapt the environment to fit the habit rather than the other way around. Contrariwise, the less frequent the recurrences, the lighter the hold of the accompanying habit, and the more amenable it is to the manipulation of reason.

The twist is that this manner of dealing with the predictable frees the always limited resources of mental energy for the opposite but complementary task of coping by means of reason and memory with the risks and delights of the unpredictable. The relative fixity of one set of responses to expected changes in the environment allows relative flexibility in responses to the unexpected. But no sooner has the unexpected appeared than its potential for being moved into the category of the expected is assessed. If the unexpected recurs, it can be passed along for action to the general economy of habit. Once the unpredictable has been slid along into the predictable, the mental resources which have been called up can be demobilized and once again put into reserve, ready to deal with the next novelty for which there is no ready-made drill. Without the formation of habit people would be so overwhelmed by the past that they could not deal with the future. In order to survive we have to "forget" in this particular manner.

The presence and vitality of the conservative tendency should not obscure the perhaps more important fact that it also strengthens the radical tendency to deal afresh with what is fresh. William James's formulation needs to be added to: habit is not only the most precious conservative agent of society, it is also its opposite, its most precious radical agent, enabling us to pay attention to new departures; this second function may be of as much importance as the first in maintaining the integrity of the individual and of society, in extraordinary times as well as in ordinary. This is the other side of the Hume principle. Although habit faces both backward and forward, the forward inclination may be the most crucial for adaptability. Whichever way one leans on that question, here is another example of the essential part the cyclical plays in making the linear possible. Contrary to much current thinking, which holds habits in disfavor, we are creatures of innovation because we are creatures of habit.

For these reasons I believe the genetic explanation is the more convincing. This view will not be generally congenial, particularly to many of my colleagues in the social sciences. People need to deny their habits if they are to be effective. They would rather imagine that their reason is the undisputed sovereign. Yet to give a role to the genes, as I have done, does not make my approach any more determinist. I do not intend it to be determinist at all; I have been trying to stress throughout that habits can be more of an aid than a hindrance to reflective thought and free choice. Holding genes partly responsible for the prominent place of habit in human behavior does not necessarily make my approach more determinist. People could continue with the same behavior day in and day out because they have learned to do so without any genetic leaning, or because they have a genetic disposition which they have learned to make use of in a particular manner. Either way, an immense variety of habits could be generated; specific genetic triggers for specific behavior could exist alongside the more general pressures to replicate what has been done before.

The mistake is to conclude that because we are not in control of our genes they are in control of us. This assumes that it still makes sense to take sides in the old nature-nurture debate (nurture being the antideterminist side) when it no longer does, if it ever did. A genetic component to behavior can be expressed only if there is an environmental component as well, and frequently will not manifest itself at all without the right environmental trigger.[59] If habits depend on a general capacity or tendency, they can be formed only from experience. Nurture is at least as important as nature. It is because nurture matters so much in the generation of individual and social habits that it becomes proper to talk about habit as a second system of reproduction different from the first, genetic system. They are both ways of slowing down the flow of time, making a record of what has happened in a relatively lasting form and from it reproducing the past (or rather the viable interpretations of it) in living form in the present.

The two systems of reproduction are vitally different, however, in the balance they strike between the changing and the unchanging. The genetic system, relying for change on mutation and natural selection, has the greater permanence and the slower rate of change, although the rate of change for humankind could pick up again if human settlements are established on other heavenly bodies with different circadian rhythms. The slowly changing genetic system embodies the experience of past evolution over billions of years up to the time when *Homo sapiens* appeared with the second evolutionary system, the system of human habit and free will which has been under relatively fast development ever since. This second system relies much more on change and less on the unchanging. A novelty introduced today can be on the way to a small measure of permanence tomorrow. The second system is somewhat Lamarckian in relying on the inheritance of acquired characters, although these acquired characters are "inherited" from experience rather than through the genes; yet if my suppositions turn out to be justified there is a direct link with genetic evolution and hence with the Darwinian system. The second system allows for much more adaptation to changed circumstances and for a measure of conscious control over the changes. Habit is the method of avoiding the reinvention of the wheel and the axle just as the genes are the method of avoiding the reinvention of the eyes and the ears; but habit allows for the much faster improvement of the wheel than the genes did of the eyes.

By distinguishing the two systems in this way I hope to dispel or moderate objections. Whenever genes are mentioned people think of a completely fixed, wired-in mechanism without any choice or plasticity about it, like the mechanism which produces blue or brown eyes or eyes at all. But a genetic mechanism can also produce a potential such as the power of thought or the capacity for habit, which is only manifested in the course of experience. There is no choice about possessing the potential, as there is no choice about having the blue eyes or the brown eyes. But people can use the potential in an immense number of ways, releasing energy for thought and deliberation by employing the devices of habit and custom for locking away that which has worked before.

Objections could be made not so much to the drift of my argument as on the narrower grounds that, if the argument is borne out, a critical feature of social behavior will be removed from the social sciences and given to the biological sciences. I have to agree, of course, that the issue is a vital one. If the tendency to cyclical behavior can be attributed to a common genetic factor (the key being the device by which recurrences are partially locked in) it becomes possible to explain much complexity in terms of simpler things—to drill through the customs and traditions of societies, which are themselves self-replicating systems, and find basically the same cycles of recurrence in the personalities of individuals and then, to drill deeper still and find similar cycles again in the brain and in all the parts of the human body down to the millions upon millions of individual cells. From microscopic genes to macroscopic societies consisting of hundreds of millions of people, and from there to the billions who compose the global species as a whole, in their social constitutions and their organic architecture the same basic cyclical processes could be at work; and I think they are, with the processes differing in their scales of magnification. But this assertion is not an attempt to undermine the social sciences so much as to add to their already large territory by proposing a potentially very considerable extension into joint territory between social scientists and those who should be their colleagues in the most momentous expedition of understanding which humankind has embarked upon. They cannot fully deserve what Henry Oldenburg said about the founders of the Royal Society (of which he was the first secretary) in the seventeenth century, but they can at least have the same aspirations: "They are followers of nature itself, and of truth, and moreover they judge that the world has not grown so old, nor our age so feeble, that nothing memorable can again be brought forth."[60]

I have tried to go a stage further in accounting for social cycles. The 24-hour circadian cycles are due, in part, to biological clocks. But, along with other cycles with periods shorter or longer than a day, they are also due to habit and its social partner, custom. Habits are cyclical because they reinstate something from the past in the present. They are mostly conservative, because they incline people to go on doing and believing what they have done and believed before. Habit also builds on solar cycles, weaving them into larger and longer-lasting persistences. My main illustration of that was the life cycle: linking one life to another, custom acts to maintain the pulsating continuity of society. Without custom, the life of any society would be short and it could be nasty and brutish. The social and biological motors, and the astronomical ones behind them, turn continuously, giving torque to society and also generating the capacity for innovation. A rotary force is imparted by the alternations between night and day, added to by the seasonal changes which have been so enthusiastically seized as markers to count by in a metronomic society which is no longer dependent on them, and which have been continuously reinforced by the cyclical disposition which underlies the cohesion of society.

FIVE

Social Evolution

Rough winds do shake the darling buds of May,
And summer's lease hath all too short a date.

Sonnet 18

Cyclical motion, with the addition of some effort and some
flexibility of response, produces a certain dynamic stability:
just as the individual organism with its members is homeorhythmic so the body politic with its members is sociorhythmic.
One kind of social cycle is in tune with biological and planetary
rhythms; other cycles, larger in number and scope than ever before, are not in harmony with natural rhythms but are still entrained by habit and custom. The second kind are apt to be in
conflict with the first. But the hold of habit and custom, even
when set against nature, is still exceedingly powerful—indeed,
social habits generated by genetic processes impart such stability
as there is to all social structures. Habits constitute the rules of
individual behavior and customs the rules of social behavior.
Although their prescriptions differ enormously between people
and between societies—perhaps I should say *still* differ—the readiness to adopt them does not.

I once asked a most discriminating and skillful surgeon who
had operated on my stomach—his skill was shown by the fact
that I was able to ask him the question—whether he would be
able to recognize my insides if I ever had the misfortune to reveal
them to him again incognito, and he what I cannot but think
would be the lesser misfortune to behold them. He answered

emphatically no. In the absence of gross pathology, a president or a prince looks much like the more humble of us, on the inside. But on the surface we are as different as our handwriting, sometimes even more exuberantly so. It is the same with habit. Just as the genetic process is basically the same in all living things, and yet produces them in fantastic variety, so the same mechanism of habit produces exuberantly different behavior. People seem to the habit born although not even an Englishman is born an Englishman. People restlessly search for what they can become accustomed to, not at peace unless they succeed; they become more easily accustomed if their model has the effortless confidence which can come only from taking the customary fully for granted. The most effective teacher is the person who does not know what he or she is doing.

The same basic proclivity prevails everywhere, witness to the obedience of people to their ruler and tribute to their common nature as well as to the kaleidoscopic invention of which they are capable. The fertility of nurture festoons itself around nature. If the tendency is invariant, it is the invariance upon which endless variations are played by those who unquestioningly wear rings in their noses or bangles on their arms, eat frogs or rarebits, take sugar or salt, chew tobacco or eschew spinach, sing for joy or grief, vent anger or hide it, walk on nails or water, speak any one of a myriad different languages or think that time runs in a line or in a circle. The same tendency lies behind the customary changes which cause those who no longer wear hats or corsets or buttoned boots or monocles, or say grace at meals, or consider Babe Ruth and William Morris idols, to laugh at the few who still do. Custom decrees that you shall be like Lermontov's hero, a man or woman of your times; the continuous remaking and reworking would be completely convincing if only people did not remember enough to realize that it could easily be themselves they are laughing at. The same universal tendency of custom constantly reduces effort by sealing out what would otherwise be demands upon consciousness.

But why, if custom is so pervasive and there is so much blind willingness to follow others, should there be so much change as well? Even if (as I have argued) habit has given us a capacity for innovation as well as for conservation, the capacity could lie largely dormant. Why should not all societies fit Hannah Arendt's description of the cyclical as "where no beginning and no end exist, and where all natural things swing in changeless, deathless repetition?"[1] Why any change, let alone the linear changes which never cease, especially in modern society, to accompany the equally ceaseless conservative oscillations?

To ask the question is to make it apparent that I have been dealing with only half the story. I have already touched on some of the discontinuities, but they were not my principal concern. I will now try to account for some of these, which go well beyond those which occur to people in the course of aging. I am going to introduce the third and last analogy from biology, from natural evolution, in order to enquire whether there is any pattern to these changes. Are there parallels, not in detail but in broad outline, between the linear developments of species and of societies—despite the great difference in time scales, with social evolution having proceeded at a far faster pace than biological evolution? About the manner in which evolution occurred in nature there has, since Darwin became canon, been a large measure of agreement. Interest in human origins existed long before the nineteenth century. Anyone born of woman is bound to be almost as interested in the origin of humanity as in his or her own beginning, both being in the mysterious past which can be recalled only by means of collective rather than personal memory.

This concern belongs fairly and squarely to the Greco-Judeo-Christian tradition, and their accounts of man's creation. Darwin and Wallace gave that same abiding interest a new twist by recounting the splendid efflorescence of life, without necessarily a hand to guide it, into the multifarious forms which have occupied one ecological niche after another on planet earth, including the development of us humans, who threaten to swamp the niches of all the other creatures. The general credibility of the story, up till our appearance, hardly has to be argued for any longer. It is the great modern linear account of our antecedents—bar Christianity, nothing has given such a boost to linearity as Darwinism—and it inevitably raises the question whether the line of development which can be traced up to the origin of human beings can be extrapolated beyond it.

It may be thoroughly egotistical, but it also seems realistic to regard the new arrival as an event second in importance only to the origination of life itself. Humans have become so much the most dominant species that the course of what has happened since our appearance has been largely settled by us. If evolution has been continuous, we certainly ushered in a new stage of it. Biologists and others who think the word is not misplaced call it the "next" stage or give it more specific descriptions like Julian Huxley's "psychosocial stage of evolution"[2] or Teilhard de Chardin's "noosphere."[3] Huxley pointed out that "all reptilian lines were blind alleys save two—one which was transformed into the birds, and another which became the mammals. Of the bird stock, all lines came to a dead end; of the mammals, all but one—the one which became man."[4] According to this view, anthropocentric though I acknowledge it to be, the future depends upon the single line which is not closed, as it has ever since it opened. It is odd that so few evolutionary biologists have as yet joined the technologists, economists, and sociologists in taking a critical look at the next stage of development.

But the question must not be begged: the very fact that the new entrant was so different may have prevented evolution from continuing. The word "next" could be used more out of the hope than the conviction that, as human beings were an advance on apes, so "modern" humans are an advance on those who went before. The gift of hope indicates one of the important differences between human beings and their predecessors, but its existence hardly clinches the argument. If hope were always well grounded it would not be needed.

The crucial quality which has given humans their advantage—their intelligence—could well have so altered evolution as to make nonsense of the continuum. Prehuman evolution (according to the Darwinian scheme) was brought about by natural selection working on and giving a direction to the random copying errors which produced genetic mutations. The small genetic "errors" which are responsible for us can be seen, with the telescopic hindsight which is presumably also not shared with other animals, not as errors at all, but as helping to confer advantages upon a species subject to natural selection in its subsequent competition and cooperation with the others.[5] Whatever else has happened, *Homo sapiens* has brought a much more obvious element of deliberation into the process of development, because we possess a large memory into which present speculations can be passed and even held while they are pondered and remade for use in the future. Any description of social development must take that fully into account, as it has to stand up to the scrutiny of the self-same intelligence.

If the only difference between human and other animals were our habits, which are as much a characteristic human possession as our intelligence, although we pride ourselves on them so very much less, or perhaps not at all, human social evolution would no doubt be much more like natural evolution. Change could move faster in the former partly because habits are passed on by copying; because many habits can be changed at any age, transmission would occur faster than at intervals of generations. But though habit is vital in social change, it is obviously not the only element. Intelligence, and the deliberation it allows, matter just as much. Intelligence produces change still faster than habit, although constrained as well as supported by it, and gives its characteristic volatility to the manner in which challenges from the natural environment are dealt with.

The environment throws up small challenges continuously, large ones occasionally; it is always and for all societies, simple or complex, in some degree problematic. It presents choices which cannot easily be closed off by being transformed into matters of custom and which are welcome to those who feel most alive when responding to the new. Some of the choices are inherently stubborn. Alfred Schutz and Thomas Luckmann give the example of a mushroom.[6] If you want to eat one safely you have to be able to distinguish it from fungi which are not edible, and it may not be possible to reduce the distinction entirely to rule, at least without excluding some possible foodstuffs. To enable the exercise of judgment, people may be forced to become as much empirical scientists as, say, the Bushmen of the Kalahari. For example, the best way to test mushrooms might be to feed them to other animals in an open-air laboratory. If the animals do not die, it could be proposed that mushrooms of a certain appearance are safe—but only until another fungus which looks the same turns out to be poisonous. Likewise, all human relationships are problematic. Constant choices have to be made according to whether the other person is kin or not, man or woman, beautiful or plain, old or young, competent or incompetent, generous or mean.

People have different heredities and are brought up differently by different parents with different resources and with different numbers and kinds of siblings. All these differences are reflected in their individual habits, even though their overall customs are the same. Tensions are inevitable and cannot always be resolved by habit alone, although all persons and societies have some habitual means of dealing with conflicts. The tension, and the stimulus, is that much greater when people in contention have different customs, that is, different constellations of habits which can be understood, up to a point, by the exercise of intelligence.

The material and human environment is inescapably problematic even when it is relatively stable; it is even more so when it is unstable. A cluster of pre-existing habits, useful in one situation, may be useless in another. The impact of people from another culture, with habits different from one's own; the rains failing; the cold or heat becoming intense; the supply of a particular food drying up; an epidemic breaking out—all put a premium on originality, which can be produced by intelligence as well as by accident. The environment changes so fast that there is an advantage in puzzlement; and finding reasons for being puzzled can become a habit of mind.

People with intelligence change the environment through innovations. Thinking about the sorts of issues that mushrooms and personality differences present, they acquire habits of innovation which introduce different forces from those of natural selection. The environment itself is then changed by deliberation. A. R. Wallace, the co-originator of the theory of evolution, pointed out that whereas other animals evolve by adapting themselves to their environment, humans with their power of thought evolve by adapting the environment to themselves.[7] This power, deriving from our special intelligence, may yet be our undoing, fostering a global rate of change which is altogether too taxing to many human beings and the environment. Our exit may be more sudden than our entrance. We have the power to impose ourselves, and if we can, we will, in small particulars like the angling of the bend on a fishhook or the manner of fitting a car into a parking space or in very much larger particulars like making a nuclear bomb.

Because we are intelligent and impulsive creatures, our capacities will out. Yet there is no obvious reason why intelligence should produce a series of changes over many millennia which are progressive and cumulative in the manner of much of natural evolution, especially because habit can draw on such deep reserves of resistance to intelligence. Habits may become more tenacious the more they are challenged, and then give way all at once to other habits no more open to reason, as happens in an apparent revolution in an individual or in a society. The transition from one set of habits to another often has to be abrupt to shake people out of their habitual (and not always benign) torpor.

Speech and Culture

The question whether intelligence has generated social evolution cannot be taken further without considering the characteristic form of its expression. Half the weight of intelligence is added by the tongue. Some other animals can talk too—whales are one of the most voluble—but not even whales have such an extensive vocabulary as most of us. The words we use create a framework of thought which is not totally ephemeral. The framework, which I shall call culture, has some durability. When we think—that is, scan the past and future and fashion the ever-moving present— we think not only in pictures, sounds, touches, and smells, which are very difficult to put into words; we also and above all think in silent speech. The words which symbolize are candidates for a longer-lasting life of their own.

When we put the words into speech, we have added something to our external environment, even if for no more than a fleeting, floating moment which hardly lasts any longer than the floating, fleeting now inside the mind. The very fact that the words have been posted outside our private envelope, even if we are mumbling only to ourselves, makes it easier to consider them as having an independent existence. They give us something new to remember. This is even more the case if the words are used to communicate with another person, for then the same words may reach port in more than one memory, and the shared memories tangle and tie each other together. If it is society which makes the crucial difference between us and other animals, it is language which makes society possible. Without language, groups bigger than bands of baboons could not have been formed to hunt, farm, manufacture, and play together; nor shared their ideas about mushrooms, stars, and gods; nor created frameworks of thought which can create a sense of kinship with others they have never seen, even lead them to expect that, if they did meet, after some initial embarrassment they would find something to talk about. Although lessons without words, and even without numbers, are always in progress, it is impossible to imagine human society if its citizens were dumb. Out of the exchange of words come the values, the conventions about ways of behaving, and the shared information which makes common life possible—constantly changing, but with its own core of stability. Information is information only because it is not immediately swept away in the torrent of the immediate present.

The next step toward durability is taken when spoken words leave their harbor in memory and enter a potentially more enduring harbor in writing or other recorded forms. Words become lodged in a space in which the eyes reinforce the ears; they can exist in a continuing present different from the present of the fabricator. As objects, words can be tested for their worthiness to be preserved, in a carol, a chorale, a contract, or a constitution. Writing not only gives precision to the collective memory; it also extends it beyond the limits of the oral and the aural. All writing, from whatever date it originated, or to whatever date it refers, exists only in the present. Cuneiform tablets are only in existence now, in the British Museum now and in books now. In that sense cuneiform tablets are as modern as computers. But because tablets, pictures, and music are on record and can be given a date, one can believe they were in existence when they were made as well as now. We can create a past and distinguish it from the present, and likewise grip the future, in a way which would not be possible without writing, and so grant both dominions their constitutions for partial independence. This illusion of the independence of the past and the future has been reinforced as the spatial environment has been filled with a forest of time tickets. Without this illusion there could be no civilization, and without the records no modern civilization.

Words have created many independent time frames and, even more stupendous, a single independent time frame for people who share a culture, complete with years and other temporal benchmarks which they number in a standard fashion. This feat is accomplished not by words alone but with the aid of anything that does not vanish on the instant. Though as much shaken by rough winds as the darling buds of May, we ourselves have some permanence; indeed we are the only thing of whose relative permanence we can be relatively sure. Even if we do not recognize in ourselves the sedimentation of habit, we can remember ourselves at previous presents and incorporate thousands of personal memories into our own personal identity. We can remember our previous memories and incorporate these too. We order memories by arranging and rearranging them in temporal sequences, according to what we remember as having happened to us, and by calibrating these private sequences with the public sequences of the general history; or, as individuals we form our own sequences, in which one segment from a childhood is followed by another segment from yesterday, or a speculation about computers in the next century is followed immediately by a "recollection" (perhaps a personal recollection of something one had half forgotten from school, which came from a half-forgotten recollection by a teacher of something he had read) of how Drake signaled to the small English ships which defeated the giant galleons of the Spanish Armada. By filing and recovering memories in chronological order we believe in our previous existence the more readily; we can confirm the order in which events occurred, because other people, older and younger than us, are engaged in the same process of ordering their memories and are only too anxious to share their reckonings for the sake of their own self-assurance.

Every artifact also serves as a memorial. Every stone that is chipped into a bowl, every lump of clay molded into a plate, every house built, represents more than a striving after permanence: we can notice that what was there is now here and predict that some of what is here will still be there in the future, and use the bowl, the plate, or the house to measure remembered and anticipated time. The same thing can be done with the natural environment, up to the most distant galaxy of which we are aware. We can measure the extent of the noncontemporaneous in the contemporaneous, and separate the one from the other. So anything which does not disappear quite as swiftly as the present enables us to reify the past and the future and mark them off on a scale of measurement; events which have happened or are going to happen can then be considered (though they are considered in the present) as distinct from events happening now.

Although a lump of clay can mold a past ("imperious Caesar turned to clay"), it is done far more extensively with words than with objects. Words above all, when repeated in their own marvelous cycles, the same and yet so splendidly different, create a partially independent symbolic world of their own. The independence of this second world means that what people say and think, does not, and does not have to, tally exactly with what they do. There has to be some correspondence; lies are disapproved of, to varying extent, in most societies. Without a correspondence between speech and action it would be impossible for people to learn from words how to act and humans would be deprived of their special educability. But the correspondence does not have to be exact. Words and deeds go together, but deeds do not necessarily have to be fully consistent with words, or words with deeds. If there were a complete disjunction between verbal "behavior" and other sorts of behavior, the gift of speech would be less of a gift than a curse to foster distrust; but it is never that complete, even for the special arts of manipulating symbols whose secrets a priestly class, or an intellectual class, has a vested interest in hugging firmly to itself. On the other hand, if there were a complete junction we would be more like very complex ants than the social animals we are, endowed with the potential of making new nests such as neither ant nor human has contemplated. If telepathic powers allowed people to communicate without speech there would be much less disjunction, but if no thoughts could be concealed, freedom could then largely disappear. It is not easy to understand why some people, looking to the future, should be so keen on proposing that enhanced telepathy would be desirable. Total honesty may produce total tragedy, as Ibsen showed in *The Wild Duck*.

The Double Helix of Society

The same point can be made another way, by likening society to the double helix structure of the DNA molecule, although once again this analogy should not be pushed too far. The two strands of symbols and behavior—or the recurring habits of mind and the recurring habits of action—would then be bound around each other. Actions that constitute behavior, if they are not automatic, are accompanied by habits of mind and other symbolic ideas that to some extent contain the information, or program, which guides the sequences of action. Often we learn the symbolic first, before it is translated into action. Each strand can be regarded as a mold for the other. They are not the same but complementary. Francis Crick says of DNA, "The two chains of DNA are like two lovers, held tightly together in an intimate embrace, but separable because however closely they fit together each has a unity which is stronger than the bonds which unite them."[8] They separate and each acts as a template for the copying of a companion chain. This is the copying process on which life is built, and it has to be done with great fidelity; even the mistakes of mutations have to be copied faithfully.

The analogy has a point to it because the strands reproduce themselves. There is also some matching between a particular set of ideas (say, about how people should behave in a family) and the actual behavior in families. The sets of ideas and the appropriate behavior that goes with them are somewhat like "rungs" on the DNA "ladder." But there are also crucial differences between the social helix and its biological model. In the former the fit between the two strands is relatively much less close. They are not so tightly bound around each other. If the fit between thought and behavior were as tight as it is between the strands of DNA, the double helix of society would be unendurably rigid. But it is not. Because the strand of ideas has partial independence from that of behavior, ideas can be developed on their own and become the means by which the other strand is observed and evaluated. There is some room for intelligence both in the way people consider their own actions and in the way intellectual resources are organized to consider the actions of the collective. Alterations are introduced not only randomly but also by the deliberate re-evaluation of cultural norms or the injection of new ideas which result in changed social behavior. The separation between the two strands introduces the additional generator of change (on top of the sometimes happy, sometimes unhappy copying errors which of course occur in both strands) which makes societies different from organisms.

A good word for the strand in the helix which comprises the symbolic (including language) is, I believe, culture, which is as essential to society as the conscious and unconscious parts of the mind are to people as individuals. The usage is not universally followed in sociology. The anthropologist E. B. Tylor, in a famous nineteenth-century formulation, regarded culture as a synonym for civilization: "Culture or civilization, taken in its wide ethnographic sense, is that complex whole which includes knowledge, belief, art, morals, law, custom and any other capabilities and habits acquired by man as a member of society."[9] The weakness of the Tylor definition from my point of view is that it equates culture with civilization, because "civilization" must include what people say as well as what they do, whereas I have been making a distinction between the two strands, between the code about how to behave—beliefs, morals, law, and so forth—and actual behavior.

This crucial distinction is brought out in a definition put forward by Alfred Kroeber and Talcott Parsons in 1958: "We suggest that it is useful to define the concept *culture* for most usages more narrowly than has generally been the case in the American anthropological tradition, restricting its reference to transmitted and created content and patterns of values, ideas, and other symbolic-meaningful systems as factors in the shaping of human behavior and the artifacts produced through behavior."[10] In this definition primacy is given to a particular component of the symbolic world, the values which it embodies. From them norms of conduct can be derived. Norms are, for Parsons, "a verbal description of the concrete course of action thus regarded as desirable, combined with an injunction to make certain future actions conform to this course. An instance of a norm is the statement: 'Soldiers should obey the orders of their commanding officers.'"[11] For Parsons the essence of culture is the set of norms which people internalize, in childhood or after, about the manner in which they should behave. These norms are prescriptive, just as habits are. Culture is made up of injunctions backed by the pressure of society (which does not mean they are always followed in practice). The view is very close to my own except that I would claim that habits are followed because they always have been, not because they can be derived from norms which can be verbalized.

To stress the prescriptive element in culture (whether it takes the form of habits or norms) is also to stress the similarity between the body of instructions contained in the genetic code and the different body of instructions contained in the cultural code. Culture, however, is made up not just of instructions but of other constituents, with varying degrees of entrenchment, which can simply be called ideas about things or values, without ever becoming instructions. One does not need to talk about their ideas to elucidate the behavior of other animals. Desert rats can be observed without wondering, or wondering too much, what is going on in their minds as they scurry around in the sand. If anything is going on, they cannot tell us. But with people it is not possible to explain anything much without paying attention to ideas—those that accompany action and those that do not. Ideas can be in a kingdom of their own even when they are not enunciated, and from there they can disappear, permanently or temporarily. Ideas are always jumping into books or computers or pots and staying there, in concealment, until they hop out again. But it is intrinsic to them that they can be enunciated, sometimes with a cruel insistence. The air is throbbing with them continuously.

The separation of powers may be the clue to the American constitution; the separation, or at least the separability, of ideas from actions is the clue to what the social sciences study: the interaction between the one kingdom and the other, or each kingdom on its own. The ideas which are common to members of a society are as necessary to them as agriculture is necessary for growing food. This is still the case now that the meaning of culture has been extended from husbanding to Bacon's "culture and manurance of mind" and from there to a particular kind of cultivation—culture as music, literature, painting, film, with the adjective "high" attached implicitly or explicitly. This last use is becoming more widespread and may drive out any special meaning given it in the social sciences, to refer to the whole code of beliefs and ideas that people possess in common.

Because culture as I have described it is made up of instructions and other ideas which have taken on the character of habits—habits of mind rather than habits expressed as behavior—culture is different from custom. In the last chapter custom referred to social habits of all kinds, whether they included action or ideas, or both together, as of course they often do. In my argument culture therefore has a narrower meaning than custom. But in another respect the meaning of culture is as wide: it refers to the whole body of habits of mind which belongs to a particular society—a Maori culture or Irish culture or Swedish culture. Culture in the singular does not mean a particular culture within the whole, a religious culture or whatever—if that is one's intent it would be clearer to talk of a subculture—but the culture which belongs to a particular society.

All four of the key words I have been trying to define—habit, custom, tradition, and culture—refer to the same phenomenon. The Hume principle, that what has been will be, applies to them all. They are all lodgements in the passage of time, crystallizations of the past which remain in the present. But I am making much of one of the four, culture, in order to stress the difference between habits of mind and other habits—even though they are established by recurring in thought just as other habits recur in action—because I do not think that otherwise human evolution can be understood. Intelligence is not just mercurial, wayward, unsettling; it is also structured in its own realm of thought. Insofar as the product of intelligence becomes part of the galaxy of ideas shared within a society it becomes part of its culture. Regarded in this light, intelligence bound into a culture has an independent

life. Culture is not just an image of a world, it is also a world of
its own with more permanence than the world of action. A language, which is a vital part of any culture, is relatively permanent
even if it is changing continuously at the margin, whereas what
is spoken with it can vanish like the breath which carries it. Most
of the language exists before any one person is born and will be
there still fully recognizable after he or she is dead. This is true
of a constitution for a state, or a body of literature, or the conventions of family life, or the rules of courtesy and other sports.
Partly because a culture has some permanence it can generate an
evolution which adds change to consistency. Its partial and continuing independence from the world of action means that culture
provides the vantage point from which society can be scanned,
and, if it can be scanned as it were from outside itself, even if
only a hair's breadth away, it can also be assessed, and judgments
made about ways in which it might be changed. The existence of
culture as something partially independent from action, or the
rest of civilization, means that some change is always being
brought about by deliberation. The culture undergoes its own
evolution, driven by an inner recurrent logic of its own, which
does not rule out deliberation and, culture and consensus not
being at all the same thing, certainly does not rule out conflict,
sometimes of the most grievous kind; the gap between culture
and what is happening to other kinds of behavior can also be a
seedbed for innovation.

Although most of what I have said about culture could apply
to a very wide spectrum of societies, it has a special twist when
applied to modern ones. I still think James was partly right when
he said that habit is a great conservative force. But a century later
the statement has to be further qualified. Habits are subject to
questioning in any society, however simple or stable. In modern
culture that questioning has become itself a set of habits which
are also anti-habits, and which many people find unsettling. Society has as it were decreed that the hold of ordinary custom
should be partially superseded by the high value placed on individual freedom. This anti-habit is as much a matter of custom as
many other habits, relying as much upon recurrence for its main-
tenance.[12] Like other habits it is conservative, but only in that it

perpetuates change by challenging more ordinary habits; and it does so from the lofty platform of abstract values which it seems almost impious to call habits at all. The anti-habit has added an ever more linear element to the basically cyclical character of culture.

Alfred North Whitehead said that "the greatest invention of the nineteenth century was the invention of the method of invention."[13] I would say that the greater invention by far has been the increasingly positive value given to the idea that individuals should think things out for themselves. Insofar as it is more than a matter of rhetoric, the anti-habit has encouraged individuals to question existing tools, ideas, and institutions. If they do so too vigorously, they bring down on their upstart heads the time-honored and often bitter disapproval of the old conservative habits, but they cannot always be vilified as outlaws from the whole value system. They pursue the novelty which is de rigueur in so many fields. In Western society freedom (since Rousseau) has become compulsory.[14]

Where resistance to this anti-habit is not too strong, an unusual system of authority can be created. Orthodox authority insists that people should continue to do as they have done before, following the same cycles and even fawning on them. The unorthodox insists that people should not continue to do as they have done, but strike out into the new, under pain of the same severe sanctions if they do not conform as those visited upon any nonconformity. An example is the youth subculture, which has some independence from the main culture because its transmittal is more horizontal and less vertical than the main culture. For those in its hold time immemorial is only ten years ago. They have left the control of the adult culture and adult authority of their parents, and although all the more subject to the authority of each other, not yet adopted the adult culture themselves. The subculture's power seems to stem not from the stability of the canons underlying it but from their instability.

The main feature of this subculture is its never-ending changes, or the sharp variations on its common themes. You have to pay continuous attention to be sure you are not caught flat-footed with the wrong shoes on. You have to wear the same style of shoes or the same track suit top as 75 percent of the other nine-to seventeen-year-olds in the country if you are to express your individuality, and finding out what your individuality demands can be a scary business. There is a kind of institutionalized charisma at work—not personal charisma (although that is part of it), because no one is sure what the source of change is. The mystery adds to the authority. Bismarck once said he wasn't the architect of German unification; he was the servant of God, and the best he could do was try and grab hold of God's coattails as he flashed through history. In a less exalted domain, the leaders of fashion are the same, like the public relations officers and managing directors of large companies who delight in telling you of the mistakes they have made with new products which their teenage consumers did not want. They are desperately waiting to grab some other god's coattails, if only they can be sure they have got the right god; one of the few pointers is the perambulating nostalgia which directs attention to periods which at least have the solidity of being in the past. It is mostly *not* the case that time past is out of date and therefore out of interest. Take this report, dated early summer 1985, London:

> Matadors are in. So is 1958. The two most fashionable looks at the moment are the bolero bullfighter look and the early beat look. Remember Charlton Heston's outfit in Orson Welles' *A Touch of Evil*? Well that was made in 1958 and that is the look *par excellence*. The question is, how do you know?
>
> If you walk down South Molton Street this week you can see all the fashion shops desperately trying to get this look right. They've got the year OK, because *Absolute Beginners* is set in 1958 and everyone knows that, but they're having a lot of trouble with the country. Bazaar, displaying the latest Gaultier collection, has settled firmly on Spain. Spain is a good bet. Barcelona was where all the paparazzis went on holiday this year, and Gaultier is usually right about that sort of thing. The only trouble is that another shop two doors down is betting its bottom dollar on Peru. Ice, down in St. Christopher's Place, reckons it's Mexico. *A Touch of Evil* was set in Mexico and it's the location for the next World Cup so they just might be right. The question is, how do you find out?[15]

You don't. But the buildup of uncertainty makes it all the easier to welcome as binding the decision once it is made (the more so because it is so short-lived). If the decision is in favor of a Spanish look then that is what it will be, until a new ordinance is promulgated, perhaps on a worldwide scale. Many fashions, in clothes, hairstyles, or music, are international, sweeping across almost all age groups in a number of countries almost at once, unencumbered by vertically transmitted local customs. What matters is not whether they are changes for the better but merely whether they are changes that have become popular. All everyone wants ardently to do is follow everyone else.

Law and Cultural Change

I must come back to the question whether the differences between human and other beings—our intelligence, our speech, our antihabits, woven together in our culture as I claim they are—are so crucial that the analogy between social and natural evolution has to be abandoned. I do not think it does. I want to claim that cultural and genetic change operate in a similar way, and that it is as legitimate to call the former evolutionary as the latter. The instructions in the cultural code are less mandatory but, even though cyclical changes predominate in both, each allows for linear change as well. Nothing I have said in this chapter is meant to be incompatible with what I said in the last about the power of ordinary habits. Actions go on being performed, and ideas go on being harbored or expressed, in accordance with the Hume principle. In them, and through them, the past lives on in the present. Unless new actions and ideas (whether propelled by antihabits or not) come to terms with past experience, including the premium which has been placed on innovation in certain spheres, neither actions nor ideas will be reproduced; but since they do come to terms, societies not only change constantly but also stay much the same from one day to another, from one year to another, from one regime to another.[16]

In both the cultural and the genetic systems nearly all changes are at the margin. They are incremental, building on what is already there. Every proposal for an innovation in behavior not only has to compete with many other proposals, most of which are discarded, but also has to be compatible with previous behavior. If the new is to be adopted it has to gain acceptance, in and out of car factories, by being kneaded in with what was already there, capturing for the new some of the legitimacy that formerly attached to the old, and maybe adding to that of the old in the process.

My specific examples will be subcultures attaching to particular social institutions, the first being the law, the second science, and the third the British monarchy. The first two were chosen from different points on a spectrum with the most conservative institutions, such as law, at one end and the least conservative, such as science, at the other. Anglo-American common law is perhaps an extreme within an extreme. Customary law is another name for it, and its content at any one period is to a high degree the same as in a previous period. The recurrences are always marked, even though less so in the United States and other countries which have inherited the same tradition than in Britain itself, and altogether less marked than they used to be. Basic traditions have persisted in good part because law is poor or no law unless it is assented to, and to gain assent it must conform to the fundamental sort of human "reasoning" described by Hume. The recurrences embedded in the law are broadly in harmony with the recurrences of everyday life and the thinking which lies behind them. It can be said that the English common law is "'immemorial' custom which ran to a time whereof the memory of man runneth not to the contrary" only because the mode of legitimacy is the same in the courts as it is outside them.[17]

This overtly customary character is no doubt responsible for the extent to which some of the salient practices have persisted, making it the oldest national law. One of King Harold's soldiers, if transported out of that other long march from victory at Stamford Bridge to defeat at Hastings in 1066 and straight into a modern courtroom, would still recognize where he was. The topography is sufficiently the same: there is a judge in uniform sitting elevated above the other actors in the well of the court. There are two parties sitting separate from each other and arguing for adjudication from on high. There are witnesses subject to rules of evidence. There is a jury of twelve good men (and now women). There are precedents for this court in the practice of the victors at Hastings as well as of the vanquished.

In the canons of legal reasoning the most vital element is precedent, which attaches authority to the repetition of an idea precisely because it is not a new one. Whenever possible, a precedent is found, preferably one which has been kept green by having been followed over and over after it was established. The respect given to precedent, which could as soon be called custom or tradition, may be more obvious in England but is by no means confined to it. Not even England is an island. A point that Hume could have made, but in context did not need to, is that to follow a precedent makes for consistency, and consistency is nine points of fairness. We are all inclined to think (as though fearing sibling jealousy like a fork-tongued devil) that if we are treated as other people in similar circumstances have been treated in the past that is "fair," whereas if we are treated differently or less favorably, that is grossly the opposite.

As for the United States, no less an authority than a former U.S. attorney general (the principal legal officer of the state, who has kept the same title in both countries) has said that in the actual practice followed by judges there is little difference between English and American law (when no constitutional issue is involved) in the extent to which precedent is observed. The respect for the doctrine is the same in both countries: "The basic pattern of legal reasoning is reasoning by example. It is reasoning from case to case. It is a three-step process described by the doctrine of precedent in which a proposition descriptive of the first case is made into a rule of law and then applied to a next similar situation. The steps are these: similarity is seen between cases; next the rule of law inherent in the first case is announced; then the rule of law is made applicable to the second case."[18]

I may appear to be arguing the impossible (beyond even the capacity of a lawyer) when I submit that the mental processes of lawyers are the same as those of ordinary mortals but nevertheless that is what the doctrine of precedent seems to show. But lawyers are also proud of their capacity for blending innovation with precedent. Oliver Wendell Holmes, writing a century ago on common law, said that although the law depended upon its past it also had to be continuously sensitive to the present:

It is something to show that the consistency of a system requires a particular result, but it is not all. The life of the law has not been logic: it has been experience. The felt necessities of the time, the prevalent moral and political theories, intuitions of public policy, avowed or unconscious, even the prejudices which judges share with their fellow-men, have had a good deal more to do than the syllogism in determining the rules by which men should be governed. The law embodies the story of a nation's development through many centuries, and it cannot be dealt with as if it contained only the axioms and corollaries of a book of mathematics. In order to know what it is, we must know what it has been, and what it tends to become. We must alternatively consult history and existing theories of legislation. But the most difficult labor will be to understand the combination of the two into new products at every stage. The substance of the law at any given time pretty nearly corresponds, so far as it goes, with what is then understood to be convenient; but its form and machinery, and the degree to which it is able to work out desired results, depend very much upon its past.

The body of law—made up of decisions which in the form of precedents impel some legal force forward into the future—can be considered the main part of its culture. The ideas on which the precedents rest have a measure of independence, however. They develop according to some inner logic of their own. Holmes also described this process well:

A very common phenomenon and one very familiar to the student of history, is this. The customs, beliefs, or needs of a primitive time establish a rule or a formula. In the course of centuries the custom, belief or necessity disappears, but the rule remains. The reason which gave rise to the rule has been forgotten, and ingenious minds set themselves to inquire how it is to be accounted for. Some ground of policy is thought of, which seems to explain it and to reconcile it with the present state of things; and then the rule adapts itself to the new reasons which have been found for it, and enters on a new career. The old form receives a new content, and in time even the form modifies itself to fit the meaning which it has received.[19]

Many examples could be given of how a legal doctrine can change, although always leaning on precedent, even without legislation, which is characteristically more linear than common law, and even in a country with a Supreme Court, which is less bound by precedent than any English court. One example is the U.S. Supreme Court's decision in *Brown v. Board of Education* (1954), arguably its most important decision of the century because it ushered in school desegregation and gave impetus to the civil-rights movement at a critical period in American history.[20] The matter was first canvassed as long as twenty years before, when it was realized that without a body of precedent there was no hope of getting a favorable decision. The old rule of which Holmes spoke did not adapt itself; it was pushed into a new form by a long-drawn-out legal campaign that did not tackle the affair head-on, for that would have been hopeless, but engaged in guerrilla skirmishes. The plan was to establish first one and then another small precedent in a series of cases, and then to parlay them all into a larger precedent which eventually acted as a precedent for a still larger decision. This legal campaign is a model of how, always building on tradition, substantial change can be brought about.

The Least Conservative Institution

The law has an essentially conservative function; it follows a generally accepted form of reasoning felt to be right and fair by any twelve or twelve hundred good men and women. Science, however, is the servant of knowledge, not justice. It does not rely on past precedent, but challenges it in the great cause of the advancement of knowledge. The new makes the old out-of-date in science as it does not in the arts. Bartok does not outdate Beethoven. Science is a force for innovation. Do not our modern culture and civilization (I now follow Tylor in coupling the two) stem from this scientific endeavor and its resonant echo in tech-nology? Science, unlike other institutions, is expected to be rev-olutionary. Granted that each scientist has an interest in the status quo "to the extent that he does not want the skills and expert knowledge which he has learned at great cost in time and energy to become obsolete," the institutional pressures toward change are still strong and persistent.[21] In this institution there is no organized conservative opposition.

This conventional view of science is itself open to challenge. One proponent of another view is Thomas Kuhn, a historian of science and the author of *The Structure of Scientific Revolutions*. His work is certainly not universally accepted, especially among those scientists who regard themselves as much less "conservative" than he is apt to regard them; he has also modified his position considerably.[22] Scientists are ordinarily more independent of the laity than lawyers; they seek agreement only from their peers in their own scientific circle. But Kuhn concludes that they also to some extent depend on precedent. John Ziman compares citations in scientific papers to legal precedents:

> The corporate, co-operative nature of scientific argument is made very obvious by the systematic use of *references* or *citations* in scientific papers. It is almost impossible to write or get published on a scientific theme without noting explicitly all relevant preceding work by other scholars . . . A scientific paper does not stand alone; it is embedded in the "literature" of the subject. Every argument that is presented, many of the facts that are adduced, must be supported by documentation, almost like the "precedents" of a Common Law judgement.[23]

Instead of precedent Kuhn uses the word "paradigm":

> In its established usage, a paradigm is an accepted model or pattern, and that aspect of its meaning has enabled me, lacking a better word, to appropriate "paradigm" here. But it will shortly be clear that the sense of "model" and "pattern" that permits the appropriation is not quite the one usual in defining "paradigm." In grammar, for example, "*amo, amas, amat*" is a paradigm because it displays the pattern to be used in conjugating a large number of other Latin verbs, e.g. in producing "*laudo, laudas, laudat.*" In this standard application, the paradigm functions by permitting the replication of examples any one of which could in principle serve to replace it. In a science, on the other hand, a paradigm is rarely an object for replication. Instead, like an accepted judicial decision in the common law, it is an object for further articulation and specification under new or more stringent conditions.[24]

A shared paradigm among scientists implies a commitment to the same rules and standards for scientific procedure. Kuhn calls science operating with a shared paradigm "normal science." In normal science the tasks of the scientist are to solve puzzles; the efforts of scientists are concentrated on particular problems. "One of the things a scientific community acquires with a paradigm is a criterion for choosing problems that, while the paradigm is taken for granted, can be assumed to have solutions."[25]

It is the effectiveness of a paradigm which makes its failure all the more obvious. Normal science does not aim to produce novelties of fact and theory. In Kuhn's words: "The man who is striving to solve a problem defined by existing knowledge and technique is not, however, just looking around. He knows what he wants to achieve, and he designs his instruments and directs his thoughts accordingly. Unanticipated novelty, the new discovery, can emerge only to the extent that his anticipations about nature and his instruments prove wrong."[26] The capacity to explore the world, however, does not guarantee that all results can be happily subordinated to a given paradigm. A paradigm leads the scientist to expect certain results. These expectations also permit him to recognize an anomaly:

> Without the special apparatus that is constructed mainly for anticipated functions, the results that lead ultimately to novelty could not occur. And even when the apparatus exists, novelty ordinarily emerges only for the man who, knowing *with precision* what he should expect, is able to recognize that something has gone wrong. Anomaly appears only against the background provided by the paradigm. The more precise and far-reaching that paradigm is, the more sensitive an indicator it provides of anomaly and hence of an occasion for paradigm change.[27]

The perception of a single anomaly is by no means sufficient to overthrow a paradigm. The anomaly is often difficult to digest, if it is seen at all. Joseph Priestley, though he made some of the discoveries that led to the isolation of oxygen, could not recognize it as a separate gas, but only as common air with less than its normal quantity of phlogiston, the gas released by heating red oxide of mercury.[28] When Roentgen discovered X-rays by accident, Lord Kelvin condemned them as an elaborate hoax.[29]

Kuhn in this connection refers to an experiment by social scientists. Jerome Bruner and his colleagues asked subjects to identify a series of playing cards, some of which were anomalies such as a red six of spades and a black four of hearts. On a short exposure almost all the subjects identified the anomalous cards as normal—a black four of hearts was thought to be a four of either spades or hearts. With further exposure people began to hesitate, but some never did get it right and found it all most distressing.

"I don't know what the hell it is now, not even for sure whether it's a playing card."
"My God."
"What's the matter with the symbols now? They look reversed or something."
"The spades are turned the wrong way, I think."

Postman, one of the experimenters, said that he himself found looking at the incongruous cards acutely uncomfortable.[30] Kuhn comments that the experiment was a wonderful schema for scientific discovery. The schema comes from the way the minds of most people work, using rules derived from past cycles of experience—even though we are as a result likely to be plunged into the same kind of perplexity as Priestley, caused by a lack of fit between theory and the observation of what we may not be able to avoid regarding as fact. In such circumstances we cannot "believe our eyes," any more than if we saw bright blue lettuce or shiny green meat.

Even when there is a fundamental shift in scientific paradigms, the new one is closely connected to the old. To begin with, the anomalies of the past paradigm define the most pressing problems which the new one must address. The successes of the past paradigm are also a source of continuity. As in a car factory or a court of law, innovators succeed with the aid of tradition even when they set themselves against it. The linear depends upon the cyclical. Moreover, the revolutionary character of a new paradigm does not preclude it from drawing on the past. As King Henry II combined established legal customs in a unique manner to give birth to the common law, so Newton, Darwin, and Einstein built on their predecessors' work, on the conclusions that stood the test of time and on the anomalies that challenged those that did not.

The precedent in law and the paradigm in science are thus alike in two important respects, although the function of one is normative and the other exploratory or explanatory. They establish rules through cycles of practice and they sometimes remake the rules into a new pattern. The result is a body of law which forms a body of doctrine consistent enough to be thought of as a whole; a law riddled with anomalies would be insupportable. In any branch of science the state of knowledge at any one time is also usually reasonably consistent.

In each case it is reasonable to think of an evolutionary path, with successive changes embodied one after another in cultural information rather as successive changes are embodied in genetic information. The path is not just one of broadening precedent into precedent, even in the law—the *Brown* decision was more like a jump than a step, even if the jump followed many small steps—and it is certainly not like that in science. The evolution is never straightforward, linear and simply cumulative. One paradigm is incommensurable with another, even though it contains part of a previous model. For Einstein "mass" meant something quite different than for Newton. Yet each step builds on another, and it is not too farfetched to consider an evolution in science, comparing the ancient classical era with the Neolithic, or the modern with the classical, or making finer distinctions than those.

The British Monarchy

My last example is a much more idiosyncratic institution, the British royal family, which has survived, against the trend, in a century which has witnessed the toppling of so many tsars, emperors, kings, and queens. It is the most traditional affair, whose main claim on people's allegiance in the present is that allegiance was given in the past. The allegiance is inherited by children from parents in ordinary families just as the throne itself is inherited. But the British monarchy might well have gone the way of all the others if to the mass of ritual which surrounds it new ritual which was in tune with the times had not been added at the margin.

The monarchy has had to evolve. David Cannadine has demonstrated how new traditions have been invented for it.[31] Tradition, being habit, begins the first time.

It is not new for the monarch to be popular. By Edward VII's death the king was well established in the general esteem: "Greatest sorrow England ever had / When death took away our dear old Dad," wrote one rhymester.[32] The use of broadcasting, however, has assured that popularity will continue by making the family on the throne still more personal and appealing. The first royal broadcast, made by George V at Christmas in 1932, associated royalty with an annual festival of growing importance and established a precedent from which neither he nor his granddaughter departed—nor, in due course, will his great-grandson when he becomes Charles III. And there has been a still more striking adaptation to TV than there was to radio. Coaches have been refurbished and Westminster Abbey made to sparkle partly so that on high occasions all will look fine and good. Queen Elizabeth was the first monarch to be crowned, as they have always been supposed to be, "in the sight of the people," or at any rate of so many of the people; and the people have continued to participate in such rites of passage of the royal family as the Queen's Silver Jubilee, her sixtieth birthday, the investiture of the Prince of Wales, his marriage and the birth of his heirs, or his brother's marriage. Walter Bagehot spoke about the evocative sight of a family on the throne; now all of us can watch an idealized family romance "live" in our own living rooms. Involvement in all recurrent ritual occasions—Superbowls, the World Series, Christmas, national ceremonials—is the rule spry potentates have always followed. It can sometimes engineer enchantment into even modern life.

The Evolution of Technology

I hope these three examples have illustrated the manner in which, despite and because of the hold of habit, institutions can evolve. In each case change comes not by a wholesale overthrowing of the past but by leaning against the past to gain a kind of leverage and add something new to it. In and out of institutions like these, cultural evolution (like natural evolution) is incremental by definition and in fact—as much in the car factory as in the monarchy. Many of the changes in particular institutions have been adaptations to more general changes in society, and it is these with which I am principally concerned. The theory of natural evolution I am taking as my analogy describes the development not of parts but of wholes, that is, whole species in relation to each other; to pursue the analogy I need to consider whether any trend can be picked out as having penetrated the whole of society.

For this analogy to be convincing, I must look not just for a trend but for a common factor or factors which might underlie it. The trend does not have to be smooth, with a steady slope to it; I can accept what Ernest Gellner said:

> Empirically, it seems far more natural now to view history as a succession of plateaux, interrupted by steep, near perpendicular cliffs—by the dramatic and profound transformations. There was the neolithic revolution, and there is the industrial one, and the sociology of either must be concerned primarily with *change*; but the sociology of the societies on the intervening plateau may tend to be "functionalist" and be concerned primarily with the manner in which they maintained themselves in relative stability.[33]

One trend has operated over the long period in which humans have congregated and cooperated, taking the plateaux and the cliffs together. I think, in broad agreement with a Marxist approach, that one long-term development has in fact been particularly pronounced: the elaboration of tools.[34] Technology has gradually added to the capacities of people, individually and collectively, in the manner that natural evolution gradually added capacities to the animals that preceded humans. By giving people a sense of control over their environment, technology has also encouraged them to think they can create their own future, and perhaps nothing has nourished linearity more than that.

Although there is a great deal more to social evolution than technology, it is the principal driving force behind the forms taken by this evolution. The parallel with natural evolution is close because I am treating tools as a kind of external organ, in the belief that the distinction that has to be made between internal and external endowments should not in itself invalidate the prop-

osition that the evolution of the social animal has been an extension of what went before. It has become rather common to say
that changes in technology alone hardly matter. Yet if the continuing natural evolution of humans increased the power of our
sense organs beyond their present strength, we would congratulate
Homo extra-sapiens on adding to the power of his receptors by
gaining longer sight and sharper hearing, and to the power of his
effectors by gaining greater strength in his muscles and more
dexterity in his fingers. Because the same things have happened
without natural evolution we can claim that a new form of evolution has overtaken the old one. Perhaps it would be more accurate to say that we have used our internal organs to develop
our external ones.

When our ancestors came down from the trees and landed, as
it were, on two feet rather than four, the other two limbs were
freed for a remarkable open-ended and open-handed development. The freedom to use our hands for grasping has been one
of the decisive freedoms, and the upright or orthograde posture
and the bipedal locomotion associated with it have proved essential to the career of *Homo faber*.[35] It is true that we needed more
than our hands. Chimpanzees can hurl stones at their enemies
and use sticks to dig up insects; but they do not have the same
intelligence or the gift of speech to the same exceptional degree,
and so have not made tools on a large scale, or used them to
make other tools, and those to make yet other tools, in a large
cooperative endeavor.

These unique endowments enabled humans to surpass not only
chimpanzees but other animals too. As they emerged from the
trees, they could not run nearly as fast as tigers and deer, but
eventually went some way toward making up for it by domesticating and selectively breeding dogs and horses, and turning them
into living tools. Since then they have gone many times better,
and on a stupendous scale, with the aid of the speedy machines
which surround their bodies and carry them so much faster than
any other animal. They have no fur to protect them, but have
used clothes as a kind of extra skin which can be put on or off
according to temperature or whim. They are not nearly so well
equipped for digging as badgers or moles but now have bulldozers.

They have no tails to swim with but have found out how to make boats and ships which can outdo fish in speed if not in grace; nor wings until they made their own which carry them along with more speed than birds, if without the same breathtaking beauty. The houses made by birds also have beauty, but they do not have rooms for use by night and others for use by day. Nor is human sight or hearing as acute as that of many other creatures, without the aid of instruments which confer a sort of telepathic power: with ordinary cameras, which have become our extra eyes; and thermal-sensing cameras, which have added an extra sense to our repertoire; and microphones, which have become our extra ears, we can now see and hear over the horizon, beyond Stonehenge, even up to the stones on the surface of Mars. At every stage we have outcompeted one species after another of our fellow animals, leaving only the fiercest competitor of all, ourselves. With our appendages we have added so much to the power of our limbs and the reach of our senses that we can surpass fellow creatures who started off with natural superiority; we have by other means done even more to augment the faculties in which superiority was ours from the beginning—above all, the capacity which has enabled us to escape a little from the flow of time, our memory.

Along this line of development the most powerful prosthetic device yet invented is the writing down of speech. Writing is not an artificial limb but, far more significant, an artificial memory which has enabled the past to be perpetuated on a scale that without it would have been quite impossible. There is hardly a scrap of the past, from the development of a tail to the development of a steam engine, or an idea from the past, from the speech of Prospero at the end of *The Tempest* to the special theory of relativity, which cannot now be mulled over in a debate conducted in writing, in which thousands of people can join who have never before heard of Prospero, Einstein, or each other. Cooperation has made it possible, and writing has been one of the keys to cooperation. No tool has done more. Other methods of recording speech have been invented more recently. With the aid of the telephone, the radio, and the television, speech and music have been made stentorian. A tiny whisper can now reverberate around the globe. We are in the process of doing the same for the related quality of intelligence—using computers to amplify the power of the brain.

The distinction between endosomatic instruments within the body—eyes, ears, wings, teeth, and so forth—and exosomatic ones outside the body, such as pens, hoes, plows, or radio telephones, has been made before.[36] It is a valid distinction, but once established is not all that important. Spectacles can be regarded as part of the eye; if the thousands of other extensions like them are also regarded as part of the body, it is plain that the human capacity for toolmaking has added to human capabilities. Even though the process has not been brought about by natural evolution, but by millions of deliberate decisions as well as millions of accidental conjunctions, it does seem justified to consider the process as another form of evolution. It would be considered an advance if people could pick out individual craters on the moon with unaided eyes, so why not if it can be done with the help of telescopes?[37] Indeed, the order of achievement could be the other way around. Not everyone, child or adult, would want to see with such hyperacuity every time he or she ventured out with a Rabindranath Tagore and peered up at the night sky. The better the sight, the less the need for the far sight and foresight of imagination.

Technological evolution has progressed by fits and starts.[38] But this form of evolution which has taken over from natural evolution has also been incremental, building on what was there before, like the law or science: "The harpoon could have been invented only by a people who already had the spear, and the loom only by those who were familiar with textiles made by hand."[39] Technological development has also proceeded through a mechanism similar to natural selection. Darwin's case was based on the vast discrepancy between the actual numbers of offspring of an organism which survive to breed and the potential numbers. Even with the slowest breeder of all, which he took as an example, the offspring of a single pair of elephants could potentially number fifteen million after five hundred years.[40] A good deal of variation would be possible over the whole range; it is not difficult to accept that some of the variations in the offspring would be favorable and thus more likely to be selected, in the sense that the offspring who were more fit would produce more offspring. Most of the variations, however, would not have this superior quality and would be extinguished. So it is with tools. The number of new tools proposed and fabricated has always exceeded the number that have survived. There have been as many false starts and extinctions as in natural evolution, the path forward being littered with Stanley Steamer, Edsel, and Sinclair cars that people did not want, with self-lighting cigarettes, self-heating soup, unbreakable china plates, floating soap, the iced-water treatment of stomach ulcers, or the hyperbaric oxygen that was once used for treating heart attacks or given to pregnant women in the belief that it would improve the intelligence of the fetus. There is no end to what hindsight judges as human folly: for there is no end to the human hopefulness which is so much one of our characteristics. Only selected improvements graft onto the body of technology and grow into greater complexity, forming a special set of inanimate habits which bear a resemblance to living habits.

Another similarity between natural and technological evolution is that although technology has for the most part developed gradually, the pace has sometimes speeded up, for instance in the two "revolutions" in methods of production. The first was the Neolithic. For at least 95 percent of the life span of the species people were hunters, fishers, and gatherers, competing directly with other animals for survival. Their methods were never static. The discovery of how to produce and control fire made the range of human food wider than that of any other animal. The use of stone for implements from hand axes to barbs and arrowheads gave its name to a period of human history. But nothing had more effect than the development of Neolithic agriculture, comparatively so recent, when humans started to cultivate edible plants and to breed sheep, cattle, goats, and pigs for the sake of their meat, milk, blood, and skins. The new methods enabled groups to produce more food than they could consume. Up to then it had been brutally true for everyone that neither shall you produce, neither shall you eat, as it still is for a large part of humankind. The changes in this period were so rapid compared to previous periods that "its beginning is often called the Neolithic revolution, using the term by analogy with the industrial revolution, for there are reasons for supposing that it was followed by a somewhat comparable relative increase in population."[41]

The pace has speeded up since the second revolution, the industrial, and in place of more or less independent developments of tools in separate societies, one main line of progress has become increasingly generalized. Mechanical power has been harnessed to old tools and made possible tens of thousands of new ones, which are like many human habits, created by deliberation to go on repeating themselves in a reliable cyclical manner. Factories have been fitted with what Marx called machines that spin without fingers; even a humble dental surgery is equipped with drill, examination lamp, tool sterilizer, cautery, and x-ray machine; and homes have been converted into miniature factories, with home laundries to take the place of commercial laundries, home ice-makers, tiny cold stores, the radio and television which have partially replaced the theater and cinema, the hi-fi which has partially replaced the concert hall, and the car which has partially replaced the tram, the bus, and the train. Such dubious delights have not spread evenly throughout the world, by any means, but the taste for them has gone ahead of their availability. People are acquiring the habit of machinery, especially machinery to move their bodies around, as part of the culture which is propelling technological evolution.

One sequence of technological innovations has a particularly long lineage. The improvement in transport has steadily widened the circle within which people can exchange ideas, goods, and themselves. By a succession of adaptations building upon each other, the (nonmetaphysical) wheel was invented, tracks and roads built, and rivers and seas navigated by larger and larger ships driven first by oars and then by sails and propellers. More recently, no linear development has been more marked than the manner in which the cart was supplemented and then partially replaced by the stagecoach, the stagecoach by the train, the train by the automobile, and the airplane.

This course of change would not have happened without intelligence, but it has not been selected by intelligence and resolutely pursued. A great deal of accident and opportunism has entered into the process by which horses tamed for warfare were later used for pulling vehicles, or railroads used for short distance transport were elaborated into comprehensive networks, or airplanes first conceived for sport were developed and adapted for long-distance transport. The elaboration of exosomatic instruments has been no more steady than the evolution from flagella propulsion to small, swimming invertebrates and on to ever-faster swimming fishes, swifter reptiles, and mammals; but in both cases the progress was undoubted. "Who, a bare sixty years ago, seeing Queen Victoria in her pony-drawn bath-chair, could possibly have imagined that within a single life-time ladies of comparable age and dignity would be stepping on the gas along the Pennsylvania Turnpike or cornering at fifty miles an hour on the Corniche?"[42] Huxley should see the Corniche now!

In communication, as in other spheres of technology, the development has been both cyclical and linear, one small improvement succeeding another time and time again in the "lifetime" of an artifact until a jump ahead supersedes all (or much) that has gone before, with the technology changing the way people think. If everyone now wants to be somewhere else, we know why. The trend has steadily accelerated, with immense development packed into this one century, marked by the spread of the car and truck for private and commercial transport and by the improvement of sedentary communications which do not require feet or wheels. People can sit in one place and talk on the telephone all over the world, or get information about it from their newspapers, radios, and televisions. The word "revolution" has often been used to describe what is happening in information technology and, despite all I have said about the gradualness of change, perhaps for once it is not wholly out of place. The pace of innovation is going to make what is available now look old-fashioned by 2000. Nothing is more striking about the modern world than the way it is generally taken for granted that communications should be improved, from roads all the way through to optic fibers. Here is one of the indisputable articles of faith in the encompassing culture of the world.

I have been claiming that there is a parallel between the manner of evolution in technology and in society on the one hand and in nature on the other. Small changes to existing habits have been cumulated into a subculture, which consists of the instructions about toolmaking which are handed down from one generation to another and the set of norms about its desirability, and this has been transmitted from one generation to another as a datum. An infant, unaided, would not understand how to produce even the simplest tool, but the infant does not need to; nor do most adults. They can take what they are given. If people could use technology only if they understood it they would still be using digging-sticks. But new tools are constantly becoming part of their way of life. A tool of any complexity is a kind of material habit which encapsulates in the present the past experience of countless generations, and which is further elaborated before being handed on in a modified form, to be discarded or modified again. The development is both cyclic and linear.

The term "tradition," which I have already attempted to define, can stand in for culture, if in this context it seems more appropriate. Peter Medawar prefers it:

> In man, ordinary evolution as we understand it in lower animals, endosomatic evolution, does still happen, and I could give examples of evolutionary changes that have occurred within the known history of the human race. But they are changes of a comparatively minor character, whereas the changes wrought upon human society by exosomatic evolution have been rapid and profound . . . Exosomatic "evolution" (we can still call it "systematic secular change") is mediated not by heredity but by *tradition*, by which I mean the transfer of information through non-genetic channels from one generation to the next. So here is a fundamental distinction between the Springs of Action in mice and men. Mice have no traditions— or at most very few, and of a kind that would not interest you. Mice can be propagated from generation to generation, with no loss or alteration of their mouse-like ways, by individuals which have been isolated from their parental generation from the moment of their birth. But the entire structure of human society as we know it would be destroyed in a single generation if anything of the kind were to be done with man. Tradition is, in the narrowest technical sense, a biological instrument by means of which human beings

conserve, propagate and enlarge upon those properties to which
they owe their present biological fitness and their hope of becoming
fitter still.[43]

The Course of Social Change

I must come back to the issue about social evolution raised earlier.
It would not be possible even to consider whether there has been
such evolution without going beyond technology to consider some
of its effects. These have obviously been very large, in the first
place for the manner in which people earn their livings. If natural
evolution continued endosomatically at the rate that cultural evolution has exosomatically we would have been relieved of the
need for industry. We would not have had to produce cars—and
so no Mr. M.—because we would have been able to run so fast
and so effortlessly on our own feet that we would not need them,
and fly so speedily with our arms—become wings—that airplanes
would have been superfluous; nor binoculars or telescopes because
our distance vision would have become so much better; nor telephones or radios because they would have been built into our
ears. If, in addition, our thumbs had become so much greener
that we could grow as much food as we now produce without
the paraphernalia of modern agriculture and horticulture, we
could have done without the paraphernalia too. In this new Ar-
cadia, a preindustrial vision of heaven, human beings (under another name to give full recognition to the new species of superman) would sit like gods on the grassy hillsides, not needing to
raise their voices in order to be heard in the farthest corner of the
heavenly universe, nor use anything but their will to summon
forth the most dulcet music, nor do anything other than indulging
their special quality (which they would presumably have not lost)
as the animal that can laugh at any such ironic predicament. But
as it has worked out, with the person now underneath the pile of
telescopes and the radios the same in most particulars as the
primitive hunters not far from the Caspian Sea, where the species
which succeeded the Neanderthalers may have come from, human
capacities have been added to only by unremitting toil and cease-

less effort by the great majority. So great is the toil, and so unequally shared are the benefits, that many have doubted whether they have been beneficiaries at all.

The division of labor is one form of cooperation; but it has been accompanied by a larger and more embracing form of co-operation which goes well beyond productive organizations. One special quality of the species is that its members can cooperate together over such large territories and across so many barriers, in science and the law, and inside institutions and outside them. Social evolution has become a vaster cooperative undertaking, and the vastness is one of its features. Without improvement of communications this would not have happened; the basic social group would have remained small and simple. The joining of groups and people into larger unities allows complex processes to occur.

The most signal means by which social evolution has proceeded is the temporal and spatial division of labor, producing the differentiation, specialization, and reintegration that have made the development so much like the course of natural evolution.[44] Specialized organs within organisms, or at least those which have survived the testing of natural selection, by and large work more effectively than the less specialized; and, however much we regret it, that seems also to be the case with productive and many other social organizations. What Adam Smith tells us about the division of labor no one has yet denied. Otherwise the scale and complexity of organizations would not have increased nor organizations spread across the world.

The more self-sufficient social units have had to give way. The family, the all-purpose institution which has served human beings so well for so long, has been the main loser. It had been the main productive unit until the industrial revolution, but, despite the long struggle it put up, could not hold out against the new industry: "Each stage in industrial differentiation and specialisation struck also at the family economy, disturbing customary relations between man and wife, parents and children, and differentiating more sharply between 'work' and 'life.' It was to be a full hundred years before this differentiation was to bring returns, in the form of labour-saving devices, back into the working woman's home.

Meanwhile, the family was roughly torn apart each morning by the factory bell."[45] Since then it has lost further functions. The family used to be school, miniature welfare state, hospital, provider of its own entertainment. But the professions which have sprung up so eagerly in the last century have extracted as much as they could from this last stronghold of generalism. The family is still that stronghold, however; if it ever ceases to be, the whole social edifice will collapse. The general culture can be transmitted to young children only by a nonspecialized institution that maintains thereby some essential variability.

This trend helps to bear out my analogy: the forms of life become more complex in the course of natural evolution; the forms of institution have become more complex in the course of social evolution. The new human capacities have been produced only by an ever widening mesh of interconnection between entire societies, with ever more elaborate differentiation and ever more elaborate integration within them. This social evolution, like the technological, has been the result of selection working on deliberation and accident to create an amalgam which is made all the more complex by the thousands of different cultures into which the new global culture has been inserted. To some extent every institution in every society changes in the same way as the law, science, and the monarchy—that is, by preserving the old in the course of adding the new to it, and legitimizing the new by kneading it into the old. Particular inheritances, of religion and art, family and government, woven into changes derived from the more general technological thrust, are what give variety to the thousand-and-one forms taken by social evolution. This variety has coexisted with a general unfolding, and indeed been necessary to it. If the further stretching of the same sort of evolution removed the variety it might also remove the evolution.

Perhaps the most significant point is that a certain direction has been given to social change by the same kind of process as that which drives natural evolution. Every change is added to a large existing accumulation and will ordinarily survive only if it is compatible with what was there before. With social evolution this means being compatible with the pre-existing culture. Accident and deliberation can enter into the novelty but it needs to be compounded with what has stood the test of survival. The legacy of the past is always large; the novelties are, by contrast, relatively small. "The idea of tiny changes cumulated over many steps is an immensely powerful idea, capable of explaining an enormous range of things that would be otherwise inexplicable."[46] Ears got their start as an ordinary piece of skin which became more and more complex, as did eyes. So it goes with the development of the common law or molecular biology, or the accretion of local custom and practice in a car factory or office. At any level the truly ambitious person has to be modest.

Because incremental changes are normally added on to what is already there, a continuous push is given to more complexity. In nature there has been increasing differentiation and specialization in organisms, and a corresponding development of hierarchical organization to integrate the ever more specialized functions. Organisms have developed more complex arrays of ever more differentiated organs; and so, broadly, has it been in society, with consequent dangers which are very obvious. The threat is that we shall all become too specialized and go the way of other species which fit themselves too closely to a niche and disappear when it does.

In stressing the similarities I may have given the wrong impression. For it would not be worth making very much of social evolution if it were only a continuation of the natural evolution which produced the Jack who made the house that Jack subsequently built. It *is* a continuation of natural evolution, but one that supersedes it. It has produced the creatures who have now embarked on a different sort of development. This development is far from over. No one (as far as I know) is saying that social evolution came to a stop, say in 1815 or 1945, and we are now in a static phase. It is still happening. Social evolution is open-ended. Every society, and world society insofar as it exists, has accumulated a body of culture which gives meaning to what its members perceive; and every society is continually changing that body of culture at the margin. Huxley's mammalian line is still open. In talking about social evolution we are talking not only of the past, but also of a present which is currently being metamorphosed.

The open-endedness persists because the process is not genetic but ideological. Gene-based construction may have reached some culmination in evolving the human brain, but in doing so it constructed the base for further development brought about by linking one brain to another. There need hardly be any limit to the extent to which people can share their experiences and ideas, even with people who are no longer alive. The cooperators in the common undertaking (including, as I say, the dead, whose experience is being constantly reinterpreted) are changing in every moment, partly as a result of a continual mutual interchange. Social evolution always results in something new to share, which gives a new impetus to further evolution. The dialectical interchange never ceases between the relatively permanent and the new candidates for permanence, and will continue as long as the process maintains the character of a cooperation within the species to preserve, to sift, and to shift.

The Question of Progress

It would (I hope) be quite possible to claim, on the basis of what I have said so far, that social evolution does unfold, without necessarily accepting that there has been any "progress." A recognizable "next" stage of the kind Julian Huxley and others talk about does not have to be regarded as an improvement on what has gone before. The question of progress is delicate ground; values of many different sorts taken from cultures of many different sorts come jostling in, and quite rightly, as soon as any judgment of this kind is made. But at this point I cannot leave the question hanging.

When biologists such as Huxley refer to the "next" stage of evolution they are proposing that evolution has continued (or will continue) and will result in as great a leap, eventually, as there was in the transformation of the amoeba "up" (one might say) into a human. This could not be possible if the only evolution is in the technology which I regard as the prime mover. Even the most obdurate pessimist is prepared to admit with an expressive shrug of the shoulders, as though it is beside the point, that there has been technological progress. The shrug asks the question whether it be boon or curse, or perhaps even implies the latter. Every tool used for happy domestic life has had a dark counterpart employed to undermine or literally destroy ancient communities. I did not mention teeth and nails as human faculties but they also have been monstrously augmented. Our technological powers have advanced further and faster than our good sense, wisdom, or morality about their use. We still find it so difficult to leave a technological advance as a potential, deliberately unused, that it seems as though technology is driving us rather than we driving technology.

The Enlightenment may therefore seem to have turned sour and the optimism of the eighteenth century, fueled by the discovery of a New World, may seem like the innocence of childhood. No one now would be prepared to agree openly with Gibbon's "pleasing conclusion that every age of the world has increased and still increases the real wealth, the happiness, the knowledge, and per-haps the virtue, of the human race."[47] Even those inclined to side with J. B. Bury when he describes progress as "the animating and controlling idea of western civilization" would think the idea is losing some of its animation, like an old man who shuffles along as if to lift one foot from the earth would bring him (and perhaps it) crashing down.[48] On the face of it, the prophecy of a Condor-cet, so hopeful about his species that he could pen it on the night before his execution by Dr. Guillotin's innovation, belongs to a remote age: "Nature has set no term to the perfection of human faculties ... the perfectibility of man is truly infinite; ... the progress of this perfectibility, from now on independent of any power that might wish to halt it, has no other limit than the duration of the globe upon which nature has cast us."[49]

Contrast the facts of the twentieth century—not just the specter of nuclear destruction but the disappointments of many newly emancipated states, the immense continuing worldwide poverty, the unemployment, the insults to our ecological environment. Even the growth of world population is not necessarily a sign of advance, except in sheer numbers. If life is worthwhile, even if only some of the fruits are yours, then it should be an achievement on the part of technology to have made it possible for so many more people to enjoy it. But this achievement, especially on the part of medicine, has taken on the aspect of a disaster, actual in some countries and potential everywhere else. Those already alive do not welcome any move to join them on the platform they must share, perhaps partly because it is crowded already and partly because they are unsure about the platform's security. They feel as though it will sink if it has to carry more than the few tons of intelligent humanity which already weigh more than the earth can bear.

And yet, although these are some of the components of the contemporary zeitgeist, I do not think it is possible to accept that all mankind can crow about (if still inclined to boast) is *mere* technological progress. Using a stock of tools whose secrets have been inherited, and adding a new one here or there, requires perhaps no more than the usual curiosity and willingness to produce and try out the new. But there is nothing "mere" about technological progress on the scale that has been experienced since the industrial revolution, and which is still accelerating. Technological change has depended on and contributed to the culture of which it is a part. So there can be no general view about progress without taking into account some of technology's wider repercussions.

As a result of a general improvement in communications the self-sufficient locality has gone the way of the self-sufficient family. The opening up of communications between societies challenges existing habits. Cicero saw this at work in the mercantile communities of the Mediterranean; it has become more widespread since, as the trade and interchange of the Mediterranean has been extended to all the seas and continents of the world. In these circumstances people seldom abandon their own habits without a struggle, except when the strangers' custom is advantageous and fits with what is already in place. Cities, which are connected with many other cities and their hinterlands, have institutionalized this sort of culture contact; so have countries which admit many immigrants from a variety of cultures.[50] New hybrid habits are likely to be the result not of original invention so much as of copying others. This has happened on a giant scale with industrialization. "Once by some miracle one society has become industrialised endogenously, others can become such by emulation. Torches can be lit, passed on and spread the light to others."[51]

Unification

The new, however, remains new only briefly. Habit begins the first time; innovation ceases the first time. The new is rapidly incorporated into, and often suppressed by, the relatively unchanging. New practices become institutionalized and lose the adaptability they had in a period of more rapid change. Defenses against innovation have become more elaborate with the increased pressure of novelty, especially when a range of compatible institutions is built into a society which is also a nation and a state. With each improvement in communications, small localities have been drawn together into larger localities, villages into towns, towns into cities, cities into regions, regions into countries, countries into regional and world economies. But as the geographical scope of "society" is enlarged, societies with common languages have been bound together into states and erected around themselves barriers to people, goods, and ideas. If that is all better communications had achieved, people would be almost as firmly divided as before; and one would hardly be able to claim that the world is being "unified."

Nevertheless I would claim it, on two grounds. The first is that the boundaries of nation-states, with some exceptions, no longer enclose something so distinctive. Diffusion and convergence have made the societies within the boundaries more and more alike—this despite the wide range of important differences which still exist. Ideas about their economies, to which almost all have given a leading place, have become more similar. The urge is for development, meaning economic development. The method is the harnessing of technology, and everywhere this technology has the same elements. A common technology calls for educational systems to teach people how to grow the new organs and to make use of them, and for systems of science and applied science to develop them, and always for improved systems of communication so that more people can cooperate in the division of labor and ideas. There is no society left untouched by the forces of technology, or which has not made their further harnessing into an objective. The old-fashioned empires, like the British, although ruthless enough about their own interests, to a large extent protected the customary from the full onslaught of the new technology by superimposing their power on the kinship-based societies they colonized, rather than trying to destroy the traditional forms of government root and branch.[52] The decline of the empires of indirect rule, and their partial replacement by empires of direct rule, has removed another barrier to uniformity.

The second ground is that national boundaries, which seem stout enough when confronted in the form of customs posts and immigration officials, are bypassed by the messages that flow over them and around them and climbed over by goods as well as by people. Leaders of state governments huff and puff about their intentions to reduce unemployment, to maintain defenses, to build up industry; but a great deal of it is pretense. The things they promise to do they cannot do and know they cannot do, even though they do not admit it. The power has passed away, some of it to governments more powerful, some of it to multinationals, and a good deal of it to the emerging hybrid culture of the world, as Coca-Cola is drunk in India and yoga centers set up in France; as French cooking is to be found in Argentina, while the samba is danced in Germany, at least by some older people; as Buddhist monasteries are opened in Sussex and a Holiday Inn in Lhasa; as Tennessee Williams is performed in Tokyo and No plays on Broadway.[53] Immigrants usually leave part of their hearts behind and remain divided in their sympathies, but they are not total strangers in their new land. The values which they observe, and others observe, are everywhere the values of mobility—physical mobility of people, commodities, and information, and social mobility through the class system. It is on the whole a merit for an idea or a practice or a person to travel widely and fast, even if it sometimes produces a worldwide catastrophe like AIDS or spreads a stock-market slump like that of October 1987.

Although internationalization has made many national governments into local governments, they do not ordinarily attempt to retrieve their power by retreating into isolation. Instead, to be connected to the world network of communications is regarded as a necessity everywhere; almost the first step a newly independent government takes, after assuring a supply of arms for its military and throttling its schools with an examination system, is to build an international airport with a runway capable of standing up to the giant airplanes which, whatever else they carry, are freighted with ideas—and not only in the Business Class. New states are tied into the world network through international temporal institutions.

The political scientist J. W. Burton has described the outcome of internationalization in this manner:

> Thanks to land, sea and air transport, to postal and telegraphic services, and newspapers and books, to radio and television and to tourism and migration, there is now a world society—a society that comprises peoples everywhere, who know of one another, and who in most cases trade and communicate with one another. Relationships are still restricted: not all people can read and write, only some have radio and television sets. Different languages and the absence of direct visual contacts limit communication, and sometimes personal movements and communications are hampered deliberately by authorities for financial or ideological reasons. However, most barriers are being reduced by technological developments, by the spread of education, and by popular demands for opportunities to be informed of what takes place elsewhere. No matter how different peoples are in culture, social development and material conditions, and no matter how geographically distant from each other, they have today knowledge of what is possible, and share expectations of increased welfare and more direct control over their social circumstances. This communication is now practically universal, even between people who do not read and write, and who do not travel.[54]

A World Language?

Another development not mentioned by Burton, which may turn out to be more momentous still, has followed the globalization of communications: a world language may be on the way to being established. Apart from the ancient language of gestures and facial expression, the first world language is a mute one. Mathematics is a perfectly adequate medium for the exchange of ideas at scientific congresses, when delegates stop talking, make do without ordinary speech and communicate largely on paper and board and screen. It has become an international language of the most genuinely international community the world has yet seen, that of science.

In this century a step toward a spoken and written world language has been taken with the spread of English. Now that the impetus has been taken over by the United States, with Indian English also spreading energetically, it looks as though the English language will be the most significant legacy of the first British Empire, which lasted till 1776, and the second, which lasted till the 1950s, in the way that Latin was for the Roman Empire.[55] Indeed, the language may constitute a third English Empire—an empire without a center, having a certain influence but no power. English has survived a chain of declarations of independence from the mother country. It is estimated that the number of speakers of English and its dialects has now grown to one billion, a quarter of the world's population, including those who speak it as a second language and as a foreign language.[56]

Language, as a flexible set of habits, seems to evolve at a rate orders of magnitude faster than genetic evolution; what is happening now to English is an extreme case. It is being elaborated at a great pace in its different varieties, in phrases like tiffin-carrier (Indian for a carrier for a snack) or Himalayan blunder (Indian for a grave mistake) or been-to boys (African for been-to-England) or minor wife (Southeast Asian for mistress) or a thousand others, and also, despite the disadvantage of its spelling, in hosts of idioms and proverbs.[57] From its main bases, English has penetrated deeply into literary forms (producing new variants of the novel, the essay, and poetry) and deeply into the language of countries which never belonged to the British Empire. Over two thousand English expressions have been recorded in French: le weekend, le sandwich, le parking, le camping, le smoking, le gangster, le gadget, le striptease, le baby-sitter, le bulldozer.[58]

There is hardly any talk now of Esperanto, or, if there is, much of it is in English. Esperanto has no more than a hundred thousand speakers. Mandarin Chinese is probably spoken by more people as a mother tongue than English, but it is not an international language on anything like the same scale, and many millions of Chinese are learning English through a vast Open University imported from England. Spanish and French and German are still in contention, and Russian or Japanese, Hindi or Arabic, may in the future spread as rapidly, but for the time being English is benefiting from natural selection in competition far beyond the immediate spheres of influence of the United States and Britain. English is the official language of the European Free Trade Association although none of its six member countries is English-speaking. It is gaining the day as the language of commerce, as the language of science alongside mathematics—two-thirds of all scientific papers are first published in English—as the language of travel, as the language of pop music and fruit machines, as the language of sport, as the language of aerial navigation and control, as the language of computers and, indeed, as the general language of technology.[59] When British engineers installed railways in Sicily or Argentina in the nineteenth century the local people did not have to speak English; but with information technology, language is much more central. It is ironic that a country which latterly has lagged in technology should have spawned the language which carries it and that successive British governments should have failed to see what an asset they have in it. Other countries, above all the United States, have become the main generating stations.[60] One of the forces propelling it is the vested interest of the urban bourgeoisie of Africa, Latin America, and parts of Asia who, as mirror images of the slaves who developed creoles and pidgins so that they could not be understood by their masters, separate their privileges off from the masses in their own countries by cultivating the language they share with North America and parts of Europe.

Through the English language, which has much of the history of England embodied in it and much of the recent history of North America, the quizzicalities of English-speaking history may become an important part of the common history of the world— that is, if the carrier does not break up into a number of mutually incomprehensible languages. Dialects within Britain have been to some extent mutually incomprehensible. Not many of the people who today refer to being left-handed as corrie-flug would recognize the terms for it in other parts of the one little country—cuddy-wifted, kervag, kittaghy, left-couch, or marlborough. A child who shouts "Croggies!" as a truce term in one part of the country (children, more civilized than their parents, respect anyone who wants to stop fighting; the inventor of a successful equivalent term for adults would be the greatest benefactor the world has yet known) would be lost in another part of the country where they use fainites, blobs, boosey, crease, cree, cruce, keppies, scrames, scrases, screase, scribs, scrogs, skinch, or snakes. There is also a warning in the way Latin split into many different languages, French, Italian, Spanish, Romanian. But circumstances are different this time around, and maybe the outcome will be too, even if the "indecent passion" for learning English now sweeping the world wanes: "We live in a very different world from that in which the Romance languages went their separate ways. We have easy, rapid and ubiquitous communication, electronic and otherwise. We have increasing dependence on a common technology whose development is largely in the hands of multi-national corporations. Moreover, we have a strong worldwide will to preserve intercomprehensibility in English."[61]

The diffusion has certainly not occurred as a result of deliberate decision by an international body, the League of Nations or the United Nations. The delegates and officials of the UN converse in English not just because it is one of the two official working languages. Nor has there been any authoritative codification like that established by the Académie Française or the Real Academia Española. The language, without guidance, has spread in much the same manner as a species on the way to becoming dominant makes its way into more favorable environments across a wide terrain and from there into the less favorable. A similar thing has happened with music (if it is considered as another language). Occidental music has been taken up in China, mostly with the opposition of the authorities, just as before Gorbachev popular music from the capitalist West penetrated despite opposition behind the Iron Curtain by reason of its mysterious attractive power. These changes have been crystallized into new habits. Whitehead's text is also mine: "It is a profoundly erroneous truism, repeated by all copy books, and by eminent people when they are making speeches, that we should cultivate the habit of thinking what we are doing. The precise opposite is the case. Civilization advances by extending the number of important operations which we can perform without thinking about them."[62]

One example of the new amplified speech that can now be beamed on a world scale is the BBC World Service.[63] It operates from Bush House in a district which, before Kingsway was built, was one of the worst slums in London.

Two rows of men and women face each other at a long table, their heads down most of the night, scanning news-tapes containing between them about one million words a day, which have been formed up into lines and rolled out onto paper from a battery of tape machines chattering away to each other in smaller rooms behind. The machines ring bells every now and then to draw attention to something urgent. The editors feed most of the paper straight into large bins by their sides, which are wheeled away when full of waste—the news that was not news enough. Nearby are smaller tables labeled INDIA, MIDDLE EAST, LATIN AMERICA, AFRICA, and so on, some occupied, some empty. The voice of a woman called the traffic manager sings out, from a loudspeaker high above the long table, that a correspondent from one of the world's trouble spots is about to file his story. The man from Tegucigalpa is having trouble. The voice interrupts. "We can hear a dog—is it a dog?—barking in the background." There is no news in a dog barking in Tegucigalpa. "Start again, please." Any of the men and women at the table who wishes can speak to the correspondent, preferably without the dog, in a "two-wayer" to gather further background to the news coming in from the ticker tape.

The business of the newsroom is linearity: if any bulletin had exactly the same content as on a previous day the editor would be suspected of a hoax. So also if he or she did what an admirably honest predecessor did in the early 1930s; on an uneventful day the radio announcer simply said, "There is no news tonight."[64] The editors' job is to chart the novelties occurring from minute to minute on the world stage. It has become a heresy that there should be none. It is not the cyclical which sparks and sustains their interest and that of their audience—the 120 million and more who listen to the 200 daily news programs in English and the 36 other languages used by the Service. The editors have to forecast daily to make sure their correspondents will be despatched to the right places—the authorities in Bonn are going to make another statement playing down the defection of one of their secret service agents, the OPEC Ministers will fail to reach agreement in Vienna, the Minister of Defence will parry the Parliamentary Question asking him whether U.S. soldiers will be shot dead if they attempt to launch a cruise missile from British soil against the wishes of the British government, the President will declare himself determined to reduce the U.S. budget deficit or get to the whole truth about arms for Iran.

Yet the editors hope their forecasts will prove wrong. What they crave is not the more or less humdrum but the genuinely unexpected, when a big story breaks and the adrenalin flows as the pace of work is suddenly stepped up, especially if another new news bulletin is due in a few minutes. A glance at the large clock which hangs over the room will tell them how long before they are on the air again. This is "real" news. If they can check it and get it out to the world as a scoop before anyone else that will be an added bonus, as happened, for instance, when they broadcast the news of Sadat's death on the BBC Arabic Service ten minutes before Egyptian radio.

The linear is picked out against the background of the cyclical, however. The generation and output of the news product is markedly recurrent over each span of 24 hours. The sleep-wake cycle is evident on a planetary scale. To track the world news, I recorded the length and origin of each item on the Reuters tape of world news, which is the one the BBC uses most. There were three reasonably distinct periods. Wherever it was in the world, the local "newsevening," between 6 p.m. and 2 a.m., produced the most news, especially early in the evening; the events of the day were frequently summed up then. "Newsday," between 10 a.m. and 6 p.m., produced less, and "newsnight," between 2 a.m. and 10 a.m., produced hardly any at all. Every country stopped generating news as its people went to sleep. Unless disaster or war kept people awake, night could be relied on to overtake the news. The timebelt during which newsnight occurs moves around the Earth like a tide, rotating counterclockwise as the Earth rotates clockwise.

Newsnight means that the BBC, at the edge of the Eurasian landmass, receives hardly any news from there after 2 a.m. Greenwich Mean Time, and after news from the Americas ceases early in the British morning input from the Far East and Australia makes up partially for the double loss. The return of daylight in one part of the world after another makes the news grow again, like a flower opening its petals, first in Europe and then in the Americas, round and round, day after day, but always with a delay. Each day the news shadow which sails along behind the darkness as it sweeps around the world persists long after the first light of dawn. People don't make news when they first get up.

The output of news in foreign languages also follows a regular sequence which is superimposed on the hourly cycle of news in English. The morning and evening in local time are the hours for the peak audience. The first broadcast in Spanish, at 8:30 p.m., is for the motorists in Buenos Aires who, three hours earlier in their time, are beginning to get locked into traffic jams and turning on their car radios. From there the World Service sweeps up the continent northward and westward, ending at 4 a.m. GMT in Central America, which is six hours behind London. The first broadcast to Japan is transmitted at 10 p.m. GMT, which is breakfast time in Tokyo, where traffic is flowing in the opposite direction from that in Buenos Aires, in, not out; and so it goes. The solar cycle operates in the newsroom as though the relays were playing a Maya-like game with time. As the Hindi and Urdu teams leave after the sun has risen in the Indian subcontinent, the Russian, Arabic, and Persian broadcasters hurry down the corridors to their studios to herald another dawn in another place, all of them chasing Apollo through the marches of London's night.

The journalists have to cope with their own cycles as well, and this is particularly difficult at night because their own most

extreme fatigue comes at a time when little news is being generated to keep them awake; some of them go out to an all-night
pub at Smithfield Market. The few people on permanent night
duty are better off because they do not suffer the customary dip
in performance in the middle of the night. But in other ways,
they may become less efficient, becoming "divorced from ordinary thinking," which is daytime thinking, like permanent
dwellers in the cave of night who would be dazzled by the light.
People over fifty-five can refuse night duty if they like. The
strain of night work can be too great if, as the sun is followed
by earphone around the revolving planet, its light is never seen.

The network of world communication, even if supported by a
common language, could be no more than a mechanical overlay
on the same old primitive territorialism, with more compunication
(to use one of the cautionary neologisms of the computer age)
than genuine human communication. There is plenty of evidence:
the utter disaster which was inflicted on Kampuchea; the throw-
back to medievalism in Iran; the reversion, for all the eager embrace of technology, to tribalism in some African countries; and,
above all, the manner in which the great and greater powers have
exploited the new technology of communication, backed up by
its sinister sister, the technology of weaponry, to pursue their
naked national interests with no more regard than before for any
more planetary concern. The world society has been taking shape
in addition to national societies, but it supersedes them only
partially. Whenever there is a conflict between the claims of the
larger unity and the smaller—nations or often territorial units
within them—the loyalty that is asserted is not loyalty to the
larger.

The Danger of Uniformity

I also have to admit that the unification I am claiming as one sign
of progress has a dark side. After a long period in which migration
of ideas and people has increased variety, we could be entering a
period in which the migration has gone so far that the social
environment will become progressively more uniform. The threat

is that unification will reduce differentiation within the larger wholes. Claude Lévi-Strauss has said:

> I have repeatedly emphasized that the gradual fusion of groups previously separated by geographic distance as well as by linguistic and cultural barriers has marked the end of a world: the world of human beings who, for hundreds of thousands of years, lived in small and durably separated groups, each evolving differently on both a biological and a cultural level. The upheavals unleashed by an expanding industrial civilization, and the rising speed of transportation and communication, have knocked down those barriers. At the same time, we have lost the possibilities offered by these barriers for developing and testing new genetic combinations and cultural experiences.[65]

Lévi-Strauss goes too far in saying this has already happened, but there is clearly the danger, or hope, or both, that it will. Since settled communities were formed, the differences between one community and another have led to wars and also to stimulus and to change. How much longer is there now before culture is stricken by lethargy and change slows right down? Even if there is a century left, the brio could go out of modern society and its component parts long before that. Perhaps it is already doing so. Effective intelligence may already be giving way to a set of new habits with worldwide homogeneity which have very little affinity with natural rhythms. Fly almost anywhere and the road from the airport already looks the same. The nightmare for the future is that however much you twiddle the knobs on your radio all stations will be bellowing out the same music, as you drive past another McDonald's.

Another danger, if one chooses to look at it that way, is that global unification will produce a world government. Such a government—unless dedicated only to preventing any power (including itself) from wielding modern armaments—would have a world monopoly of force. Such a government could become the most oppressive there has ever been, and strangle the heterogeneity which is as essential as the homogeneity. If there is to be unification without a government to preside over it, however, a great deal more thought is needed than the issue has had. Anarchically minded people who dream of societies without governments have imagined that they would be possible only in small territories. Utopian communities and communes without central authorities have always been on a small scale. It would be strange indeed if the first working society without a government turned out to be at the opposite extreme, on the largest scale so far imaginable. But it could happen, and indeed appears to be happening, alongside the nation-states. "Complexification"—specialization between countries and between the many new institutions which have grown up—has been accomplished within an ever larger whole. To go back on that would halt evolution just as surely as any complete global uniformity.

So, recognizing that the large can be beautiful as well as the small, what has to be striven for is what usually cannot be attained, the best of both worlds. Although Julian Huxley and the others who hoped for a next stage of evolution have not gone into much detail about it, I think it is now clear that the new "species" being developed in the next stage will take the form of a new kind of body, a global society, rather than a new kind of individual—or, if an individual, one that is able to operate on a very large as well as on a very small scale. A world society will be the new species. Its individual members will not necessarily have any superior intelligence or other qualities but the result will be the same as if they did because they will share their thinking and their work with more people and to better effect. An international lingua franca will be more essential to inter-thinking than almost anything else.

An increase in the number of people who could share their thoughts and actions, with the aid of these new communications, could still be a dead end. If all the thinkers had the same thoughts, one of them would be as good as many. There is no point to it unless they have different thoughts to share; that means more diversity, not less, within an overall unity. It means more individuality, as well as more sharing. Ideals will need to be shared as well as other ideas, and there will always be a danger of the common values freezing into a common rigidity. The human race will not become a durable society unless there is a looser fit than any society has ever before allowed between the two strands of the social helix and between the members who occupy different niches within it.

It is hard to imagine people maintaining diversity unless one form it takes continues to be geographical. The most hopeful possibility is to welcome autonomy within a world framework, but without again elevating the nation-state to sovereignty: it has been too good a home for belligerence. The world can no longer afford people willing to die for their country; such suicides could be dragging to their deaths everyone else who has not consented. Only the increasing scale of organization has fostered unification. This has often reduced autonomy at the national levels, whereas the task now is to expand it, especially wherever it promises more organizational variety. If there is going to be less diversity to cultivate from without, there needs to be more cultivation of diversity within. States could become positively benign if shorn of their sovereignty, not by any formal surrender to a United Nations but by informal surrender to the new international networks with their sometimes mysterious dynamics. I am thinking more of the new networks of culture than of mere multinational corporations. This would be more likely to occur if the dominant powers of this or some future time became so entangled in ever denser networks, and loyalties so much diffused as a result, that some of their legitimacy leaked away. A world without any puffed-up governments is difficult to imagine. It will certainly not happen just because it would accord with natural evolution—in which metamorphosis has sometimes been spurred by the coming together of species which had previously been isolated on their "Galapagos Islands." If it did happen, states which cultivated and protected variety would prevent, or at any rate slow down, the processes of homogenization.

Whether national governments ever become local governments, without any overall world council to make sure they do not rise above their station, the more diversity that can be encouraged, the better for the course of the overall complexity. On this question of how to encourage "internal" heterogeneity in every way possible, individual intelligence and the intelligence of inter-thinking groups need to be brought to bear, because only "mutations" will both join and redirect the general flow. Authority would never be quite the same again if states and nations no longer needed to be in a constant state of discipline, standing ready for war. There would be no need for authority to plant itself so firmly in our minds by enslaving our habits. Perhaps societies are beginning to loosen up, and to have their class systems called into question. Perhaps the diffusion of autonomy will follow other kinds of diffusion, or is already doing so as an ingredient of the same general evolution of humankind. Like all major social changes, this one is already fraught and will become more so. Acceptance of a world culture, and an open-ended one at that, seems to many people more like anarchy than the precondition of further evolution. Society without the authority vested in states, and the hierarchies of barons who exercise it on behalf of states and themselves, is not yet easy to imagine.

Given favorable conditions, however, society will be on the way to becoming a species of a quite new sort, which boasts of only a single specimen—at least until other heavenly bodies are colonized. Its intelligence will be even more a shared intelligence than it is in the societies of a more ordinary kind that some animals and almost all people live in now. Habits and memories will be global, and their inter-thinking organization will be on the largest scale that the planet allows. If this development meant no more than a merging of individualities, people as the carriers of immortal ideas would be confined within a lumbering world society as genes are contained within our lumbering selves. The prospect could be horrifying, particularly if a global society were matched by a global government. Social evolution could then stop, many millions of years before its potential had been realized.

It would stop because evolution would have taken one fork instead of following its two diverging paths simultaneously. Social evolution has taken the same form as natural evolution, with more complex methods of integration accompanying more complex methods of differentiation. The two always need to be in balance. Once it has become global, society will continue to evolve only as long as the individual people (as well as the individual territories) which make it up are more diverse, more original, more individual. The new species will not be doomed to extinction as long as the anti-habit can be even more fully cultivated, without destroying the social cohesion which makes freedom possible and so turning hope to dust.

It is all very well to imagine society as a body so long as it does not actually become one, with a single head. Such a body has been conjured up before. "The whole of society," said Lenin, "will have become a single office and a single factory, with equality of labour and pay."[66] That society would be a great deal worse than what we have had so far. It would also be constantly vulnerable. If a single society had a single set of habits and a single, if truly immense, information store, it could be captured by a tyranny from which there would be no escape. The temptation to aspire after that kind of power would be very great. Or (more insidious still) there could be a tyranny without a tyrant. The system could be the tyranny, with an overriding dedication to preventing change that did not emanate from itself. If human society ever reached that point, unable to evolve any further, it would surely go the way of the dinosaurs. The open line has to be kept open.

A Cultural Counterpart

Although granting that if there has been any progress it could stop forever tomorrow, and is therefore more fragile than when the world was less of a unity, I believe that the technological unification which has allowed for new forms of social diversification has not been without some general cultural counterpart. The old structures certainly persist, on the ground and in people's minds; people belong to families, local communities, nations, and no doubt think of themselves as belonging more to them than to the abstraction of the world. Nevertheless, it is impossible to ignore the fact that small and large communities, small and large nations, have been partially overrun by the crisscross recurrences over the global surface of trade, travel, education, ideas, and the media; and by worldwide urbanization, which has brought more and more people into cities which are all in some degree city-states in ready communication with other city-states. The way people regard themselves and others has changed as a result.

One clue is in Durkheim. He saw that when a wider division of labor replaced a narrower, when nations were organized into comprehensive economies in place of what had been largely self-sufficient communities, and when nations became part of a world economy, new sympathies were apt to develop. If people were less at one with those in the same territory, they were more at one with those in other territories who were also weavers or welders, scientists or serologists, journalists or gymnasts. If their sympathies were attenuated, without the heat that localized attachments can generate, they were also extended from those nearby to those far-off.[67] All that is more true now, with the mass media which are not native to any one country, not even the United States, molding new habits of thinking, feeling, and believing which are ever more wide-reaching. The warmth of love may have thinned down into tolerance, but its reach is much wider, with a world to span.

Cooperation has been steadily extended. As new habits of mind have spread, larger stores of information have been shared by more people. The common history has been opened up by the printing press and its successors. The discovery that before our present continents drifted apart they formed a huge southern continent posthumously named Gondwanaland, consisting of South America, Africa, India, Australia, and more, is as widely known as yesterday's events in the Middle East. Attempts to keep information secret (and they are ceaseless) are no longer very successful, the more common attitude outside commerce being that ideas are to be drawn on and that an inventor of an idea gains rather than loses when the idea is taken up by others. The venture most characteristic of the century has been the creation of a vast information cooperative, divided into many enormous sections. In this too there is a parallel with natural evolution; to bring it about, genes had to cooperate on a massive scale with other genes belonging to the same population in a species. Through their cooperation large gangs, and eventually bodies, were evolved, with the ten trillion cells in each human acting together as well as separately.[68] There is no more spectacular cooperative than a human body, and none more extraordinary than a human society.

To that extent the processes have been very similar. Is that also true about the part played by competition? In the case of natural evolution constant pressure has come from competition within and between species. Richard Dawkins likened it to an arms race, evidenced on the grand scale by the struggle between carnivores and herbivores, the one side improving the means of attack in an interactive mutuality with the other side's improvement of the means of defense. Cheetahs, for example, became fleeter of foot, keener of eye, sharper of tooth in order to hunt gazelles as gazelles added to the speed with which they could run away from the cheetahs. Competition has, of course, performed the same part in the propagation and selection of social conventions and still does. The analogy of the arms race is both realistic and ominous.

The value of this competition is not to be played down just because it is associated with self-centered self-advancement. Competition between different habits in and out of industry is the main check on the tenacity of their hold. If everyone had exactly the same notions about and exactly the same practices for gathering and assessing mushrooms or making telescopes or speaking English there would be no need, or spur, to change anything. The universal tendency is for any person, institution, or society to crystallize into an unchanging, inflexible structure. What keeps it from happening is constant competition between different ideas and different ways; whenever an alternative is presented, a choice is possible and freedom can be exercised in making it. If cooperation is the ally of the cyclical, competition is the ally of the linear. It is above all competition which is the selective agent for countering one habit with another and bringing more fully into the open some particular element of the culture which had previously been taken for granted.

To my mind, however, the creation of the common store of information which more and more people (whether cheetahs or gazelles) can take from and contribute to is even more characteristic of the modern age than competition. There have been many Attilas and Alarics since both cheetahs and gazelles were outdone. Even when they did not go to extremes, many people who followed Darwin (particularly those who had done well for themselves by faithfully following their own interests) stressed the competition in nature almost to the exclusion of the cooperation. They picked from Darwin what suited their own position in society as a successful competitor, regarding themselves as more like a cheetah than a gazelle. But if cooperation has become more fundamental since the appearance of *Homo sapiens* and if mutual aid in the one dominant species has been more crucial to social evolution than mutual struggle, social Lamarckism could turn out to be quite as significant for the social ethic as social Darwinism was. There has not been such an information cooperative before, with such wide-ranging cooperation between the living and the dead and between people widely separated by distance, and it cannot fail to affect not only how people behave but their views about how they should behave.

The extent to which we cooperate is concealed from us. Habits and especially their collective form, custom, rest on cooperation. They enable us to borrow so well from hundreds, thousands, millions, or billions of other people partly because we do not know what we are doing. We imagine to be our own work what is really other people's. The great structures of habit which bind the past into the present and which enclose us are also great structures of cooperation: most habits are not individual affairs but customs which we share with other people, imprinted on us from birth onward in the way we eat, in the clothes we wear, in the languages with which we communicate, in the skills which we learn (and forget) for making a living, in what we regard as assets in members of the opposite or the same sex, and perhaps above all in the concatenation of beliefs which we absorb from others and give to others to absorb. Everything we are we got from others, except at the precious margin of our individual and social continents, and in everything we show ourselves the most cooperative animal, able to hunt, quarrel, perspire, and laugh in bands of millions. The paradox of society is that people cannot know what they are without an unfolding of the onion skins of self that is beyond the capacity of anyone bound by the "law of the disappearing cycle." Cooperation is the deep structure and competition the superficial; the former sets the framework in which the latter is given play. Yet we cannot fully know that this is so and that the deep structure is always supported by feelings that we are not a series of I's but a series of we's. Instead we devote nearly all our attention not to the affinities between us and other members of the human race but to the gorgeous differences which excite our loves and our hatreds. The deep fraternity does not foster fraternity in action; instead it often allows discord to excite the actors with eagerness to win predominance over the common clay.

But we do not need to remain bound by ignorance and illusion about what we are. We could become more aware of how much we have built on our childhoods, when we remain dependent over a longer span of life than the members of other species. We could become more aware of the resulting capacity for sympathy which, when we do not shut it off, enables us to project ourselves into others. We could reveal more of the living tissue of fraternity and become more aware of the way we have used our trained capacity to spin an ever more complex web of social cooperation. We could even agree with another biologist, Ashley Montagu, that "indeed, without this principle of co-operation, of sociability and mutual aid, the progress of organic life, the improvement of the organism, and the strengthening of the species, becomes incomprehensible. Co-operation constitutes a stabilizing, a cohesive, factor insofar as it makes for successful group activity, and thus ameliorates the environment so that the members of the group function more effectively and the survival rate is increased."[69]

If we did become more aware of our mutual dependence, adults on one another as well as children on adults, that could encourage yet more cooperation, and even a more cooperative spirit, up to the scale of the species. We cannot hope, nor should we, to unravel our individual and social habits—the information overload from which we already suffer would become quite insufferable. But becoming more aware of the process of habit formation, and to some small degree seeing ourselves as though from outside, could bring on a more energetic, conscious push for cooperation. Social evolution would take a perhaps decisive step toward greater self-consciousness.

The replacement of smaller principalities by nation-states could provide a precedent for the changes in people's outlooks that would accompany such development. The nation benefited when nationalism replaced localism and a new notion of citizenship came to the fore.[70] The Gascon did not once have much affinity with the Breton, the Aberdonian with the Cornishman, the Californian with the Yankee, the Bavarian with the Prussian, the Ukrainian with the Uzbek; whereas now these complex networks of reciprocity may be tenuous, but at least there is some measure of respect or tolerance for their mutual interests. Strangers are no longer so suspect, nor can they be treated so brutally, if they come from within the same country of sentiment. That country has become larger. It is difficult to deny that there has been any change in morality to accompany material improvement when the history of punishment is invoked. Here is an account by Michel Foucault of how punishment was once employed. On 2 March 1757 the regicide Damiens was condemned

> to make the *amende honorable* before the main door of the Church of Paris . . . wearing nothing but a shirt, holding a torch of burning wax weighing two pounds . . . the flesh will be torn from his breasts, arms, thighs and calves with red-hot pincers, his right hand, holding the knife with which he committed the said parricide, burnt with sulphur, and, on those places where the flesh will be torn away, poured molten lead, boiling oil, burning resin, wax and sulphur melted together and then his body drawn and quartered by four horses and his limbs and body consumed by fire, reduced to ashes and his ashes thrown to the winds.[71]

The quantitative arts of torture were cultivated so as to graduate the criminal through a thousand deaths as a public spectacle, with every scream and every movement of his head noted by the multitude as they and the criminal watched his body being torn apart. The sceptic can assert that the arts of torture are still cherished. He or she may be little moved by the fact that the torturers may perhaps be ashamed of what they do in private instead of proud of what they did in public, little impressed by the fact that there was no Amnesty International in eighteenth-century France, or given much pause by the fact that in many nations, with the progress of medicine, more art and science now go into the reduction of pain (and the collapsing of a thousand deaths into one) than ever went into the cultivation of agony.

Foucault apart, I think one still has to admit that the case for progress—which is very strong if the memory is cast back a few thousand years—becomes a little weaker as the "time machine" of the present enters the last few hundred years, and weaker still over the last few decades. Perhaps this has always been so. The tendency in any period is for people to believe things have been getting worse, to romanticize the past, to claim that there has been a decline—perhaps due to identification of the past in general with one's own childhood. The nearer the approach to the present the more solid, it always seems, is the ground for pessimism; and it is always difficult to accept that though we live in bad times that is nothing new, and perhaps the bad times of the present may not be as bad as those of the past. For people only too conscious of the shortness of their lives it is not much consolation to say that they are being too impatient in demanding visible evidence of progress in the short term. Cultural evolution, even though it works a great deal more rapidly than nature's, still does not move fast enough to be perceived within the span of a lifetime. Moreover, evolution of any kind can have a temporary halt before resuming its advance. There can be breaks in the linearity.

Whether there has been a break, like the main question of whether one accepts that there has been progress in social as well as in natural evolution, must be a matter of personal judgment. I do not doubt that exosomatic progress has continued, and been speeded up. I do not doubt that there has been further differentiation within more global hierarchies, nor that it has been accompanied by the growth of world-mindedness that would have commended itself to Condorcet. Even though its growth is still as much potential as actual I do not believe that a starving child is still so much thought of as a member of another group for which one has no responsibility. I also do not doubt that destructiveness has been enormously and hideously augmented by the same evolutionary thrust, and that sooner or later warring states, and warring factions, may make use of the destructiveness on an unprecedented scale. There has been a bifurcation in the continuing evolution which never happened while weapons were more like ordinary tools, with one array of the expanded powers by and large benevolent and the other malevolent. If I plump for a balance which gives more weight to the benevolent than to the malevolent I have to recognize that this is no more than my opinion, and the opinion is partly based on the belief (which has yet to be proved) that nuclear arms are a common world issue which could in time pull people together instead of pushing them apart.

Important as it is to consider whether evolution has been continued in human organization, a still more important issue is whether in the present estate and order of mankind we should still aim at progress. We can join with Nadezhda Mandelstam, the wife of the great Russian poet, when she said in *Hope Abandoned*, a volume of her extraordinary memoir, that whatever the future held, she only hoped she did not live to see it with her mortal eyes.[72] Even she had not quite abandoned every vestige of hope. In going so far she was highly unusual. Hope is for most of us too much a habit of mind to be abandoned. Can any of us look at a child (especially one's own) without hoping that it will grow up a little further from imperfection than its parents? It is in the spirit which has driven human effort so far to believe that life will continue to be more than a tale told by a randomizing idiot; to consider the world as more than "this great stage of fools"; to reject the notion that time produces inevitable deterioration; to harbor, in E. H. Carr's words, "a belief that we have come from somewhere . . . closely linked with the belief that we are going somewhere"; to recognize that human beings are still driven by the insistent need to surpass themselves; to share the assumption of science that the greater and still-hidden truth is ahead of us; and to agree with Oscar Wilde that "a map of the world that does not include Utopia is not worth even glancing at." A map of the universe should now be added to the map of the world.

The evolutionary scheme can be made even more sweeping. J. T. Fraser, the theorist of time, has distinguished four "stable integrative levels" before the further two represented by human beings and the institutions of human society. They are radiation (the world of particles with zero rest mass); subatomic and atomic particles; discrete, massive aggregates of matter (the universe of galaxies); and living organisms before man. "Each of these integrative levels subsumes a rich, hierarchical order of structures within the domain."[73] To extend the scheme in this way only makes more pointed the question I have been asking, as I have moved away from the cyclical to consider the linear, about the last stage in that series. Can one legitimately believe that there has been any social evolution to parallel and continue on from the others? I started off by arguing that social change is the unavoidable accompaniment of habits, above all because the culture of a society fits loosely rather than tightly with the rest of behavior. The gap between the two, leaving a good deal of room for innovation, has become wider in modern society. Technical and social change has been speeded up, but can it be considered by analogy as evolution? My view is that it can. Technological evolution is undoubted, and social evolution almost as much so. Whether they have brought about any similarly sweeping "progress" (the most linear of linear concepts) of a more general sort must be a matter of opinion. However unfashionable it may be, my own view is that what has happened represents progress even if it has not always been clear whether any two steps forward have been larger than one step backward.

The question has been pungent only since the Enlightenment; every age thereafter has had to find an answer appropriate to itself. Finding one is bound to remain part of the task of cultural re-creation which no society can avoid. As the genes were nature's greatest achievement, so culture has been the greatest human achievement, all the more so if it can be questioned. The enterprise sketched in this chapter will continue as long as human life does, and it will continue to involve the whole of humankind, with increasing intercommunication across divides which will become more and more surmountable. Innovators in thousands of different fields, bound to the more dynamic habits which fit our age, will go on proposing novelties and putting them into practice. But it is not they who will decide whether their own particular novelties are more than a nine-day, or nine-month, nine-year, or ninety-year, wonder. The members of the species will decide, not like a jury of the world awarding a verdict by referendum to one of the scintillating advocates at the bar, but like a democracy that votes by its actions rather than its ballots. In that great worldwide assembly which is in permanent session most new proposals, however brilliant, will disappear—although seldom without trace—while the occasional cluster of innovations, worked and reworked, will enter and reenter the ongoing stream of culture which is the common heritage of the world; and there remain. The life and death cycles of ideas will continue to seem to be governed by accident, and continue not to be.

The Metronomic Pulse

And then he drew a dial from his poke,
And, looking on it with lack-luster eye,
Says very wisely, "It is ten o'clock.
Thus we may see," quoth he, "how the world wags.
'Tis but an hour ago since it was nine;
And after one hour more 'twill be eleven."

As You Like It, 2.7.20–25

Economics, not sociology, has been called the dismal science. Nevertheless it may seem odd to propose from the point of view of a sociologist not only that social evolution has superseded natural evolution but that the old doctrine of progress does not have to be discarded from either sort of evolution. We are not exactly in a golden age of optimism; on these matters I count myself a contingent optimist, which is what all but the most sunny optimists have to be.

Insofar as there has been a linear evolution in society it has introduced new cycles appropriate to an industrial society, and whole new practices in the management of time have been introduced into each strand of the social helix. The advance of technology, and the two-way influence between it and social institutions, has depended on advances in measurement. In the culture and practice of the metronomic society everything measurable has been measured: space, objects, intelligence, performance, and almost everything else in terms of the ubiquitous counting device of money. Above all, time has been measured. Without the new customs about time which have been developed over the last two thousand years and more, either there would have been no social evolution or it would have been quite different. Unless we had pinned it down (or tried to) with the artifice of calculation, with a precision not wholly spurious, there would have been little or no growth in technology. Without taking stock of what has been done to the measurement of time, and how, it will not be possible to take into account the "hurry sickness" which now seems to be endemic in modern society and which would, no doubt, have seemed a strange outcome to Condorcet and his fellow optimists.

There is nothing new in time measurement. Nature has its own marvelously intricate methods, not all of which can be put on one side. The earth and its inhabitants still need rest. The genes are hitched to the sun. But people have also invented their own new systems, developed by evolutionary steps from what was there before. People knew how to tell time before they had any technology to guide them. The motions of the earth saved them from having to confront a virtual infinity of time. Without these markers, there would be nothing between a person and the terror of the boundless. We have been preserved from that terror and our environment has been charmed into a certain homeliness by the series of moving boundary posts that the rotations of earth and moon put between us and the cosmos. The regularities of the year, the month, and the day are the triple social security which gives people the confidence to cast their imaginations back to the origins of the universe, to the beginnings of our planet, and to the early us on it, and forward to our various possible ends. We can reach out to the large from the small. We have the security of counting our beads both forward and backward instead of gazing out from a motionless platform faced by space without time, or at least without the movements in time to which we have become accustomed with the very genes of our bodies. These natural regularities enabled us to make a whole series of collective decisions about how to improve on natural timing, which have eventually been embodied in a culture with an increasing geographical sweep.

Agreement on time measurement was necessary before larger numbers of people could increase the scale of their social organization. The first need was consensus about when a year was thought to begin. It would be intolerable for people adding up the days in a year to start from many different zeros. Our own calendrical system derives from the Roman. The year in ancient Rome began in March (September was originally the seventh month, and so on with October, November, and December). Before his name was given to the month which had up to then been called Quintilis, and the following month was named after his successor, Augustus, Julius Caesar decreed that according to his Julian calendar the year should begin on the first of January. His decision has been truly imperial: it has been obeyed over a vast space and time. Caesar chose the first of January because this day saw the first new moon after the winter solstice, and so could be thought of as a double beginning of the year, blessed evenhandedly by both the great luminaries. In England and Wales the legal year that was established in the middle of the twelfth century began on 25 March, at the Feast of the Annunciation, although other calendars continued the old Roman attachment to 1 January. The start of the tax and general financial year was shifted to 5 April in 1752. An 11-day correction was necessary when in that year the country changed over from the Julian to the Gregorian calendar. The violence done to the calendar caused considerable unrest: there were riots in the streets of London, and workers demanded to be paid for the days they had lost.[1] That variation on tradition, like most of the traditions about time measurement, can be precisely dated.

Another decisive move toward linearity was the acceptance of another equally arbitrary decision, about the beginning of a series of years. Any year can be taken as the beginning, but to put a frame around some number of years which will be "ours," a starting date with some special significance has to be chosen. Though the main world religions have all chosen to start their counting systems from the births or deaths of their founders, other systems have flourished as well.

> There have been introduced such social frames of reference as the death of Alexander or the battle of Geza among the Babylonians, the Olympiads among the Greeks, the founding of Rome (*anno urbis conditae*) and the battle of Actium among the Romans, the persecution of Diocletian and the birth of Christ among the Christians, the mythological founding of the Japanese Empire by Jimmu Tenno and the discovery of copper (Wado era) in Japan, the Hegira among the Mohammedans, the event of the white pheasant having been presented to the Japanese emperor (Hakuchi era).[2]

As in most instances, in the Christian case the decision was not finally made until long after the founder's lifetime. In the sixth century an abbot, Dionysus Exiguus, proposed that the Christian era should date from a unique event of far-reaching religious significance, the supposed year of Christ's birth. The abbot has proved worthy to join Caesar in the temple of the most successful reformers ever; he helped to linearize the way people think, with the new age inaugurated then destined to continue until "the Second Coming of Christ would mark the end of the world, which would be tantamount to the end of time."[3] The abbot's proposal has gained increasing currency; now, when 1988 or 1991 is a stock notation in most places, the Christian number, at least in the calendars of business houses, is often put where necessary alongside the Muslim, Buddhist, or other number. A calendar I have from the Saudi Cable Company has the double heading of 1987 and Hegira 1407. The Christian has become a kind of "Esperanto" calendar.

The standardization spread when the Greenwich Meridian was established as the zero longitude. Time was in the special care of the Royal Observatory at Greenwich from its foundation in 1675. David Landes in his great history of clocks has described how John Harrison, a carpenter from Barrow-on-Humber, built the first accurate marine chronometer that would stand up to the harsh conditions of a sea voyage. He thereby qualified for a prize of twenty thousand pounds offered by the Board of Longitude. Harrison worked for 40 years on the instrument. On a trial voyage from Portsmouth to Barbados in 1764 his chronometer lost only 39 seconds. Harrison wrote: "I think I may make bold to say that there is neither any other Mechanical or Mathematical thing in the World that is more beautiful or curious in texture than this my watch or Timekeeper for the Longitude . . . and I heartily thank Almighty God that I have lived so long, as in some measure to complete it."[4] Such enterprise helped to make Greenwich the standard reference point for navigators long before the International Meridian Conference in 1884 gave it an official stamp, against the trenchant opposition of France. When the earth was partitioned into time zones later on, Greenwich was the ready-made benchmark.

Division of the Day

The passage of years can be counted without too much dispute; the position of the sun is decisive. Lunar months can also be linear months: before electricity the moon was second only to the sun as a celestial clock and a source of light. Its apparent motion is not quite so even as the sun's, but it changes its visible face every night, it moves faster across the sky than any other heavenly body, it denotes an indisputable beginning and end to each lunar month, and its movements stir the oceans into an awesome and awesomely regular tumult. Months can be counted as easily as days. In each case the cycles can be added up and used for linear calculation.

Divisions within the day were much more troublesome. It could not, until fairly recently, be conveniently converted into a segmented line like the longer spans of years, centuries, millennia. There first had to be a decision about the intervals to use. In our custom the Babylonian sexagesimal system of fractions provided the framework. It produced 12 hours in a day and a night, 60 minutes in an hour, and 60 seconds in a minute. It was a theoretical computation, to begin with—there was not yet a way of measuring such fine divisions—useful as a hypothetical structure, even though for many centuries the hours were not the same length. For the Romans, and others who came after them, the night always had 12 hours, which grew longer each night until the winter solstice, and shorter until the summer. No more was possible until measurements could be made within the night; before artificial light the night hours could not be used for much that depended on knowing the time. Not until clocks were invented, in about 1330, did the hour become the modern standard hour.[5]

The inability to measure time precisely within the day or the night ruled out the sort of 24-hour synchronization that is essential to make the best use of space and time in a technological society.[6] Despite its gradualness, dawn was therefore a crucial signal. When large numbers of people had to be brought into unison, for the start of a battle or a public hanging, dawn was a favorite hour, with the sun telling the time—sometimes aided by

the roosters that Landes tells us English soldiers took into battle
to avoid being overrun in the half-light while they were still
asleep.[7] Before the clock, large numbers of people could not be
synchronized unless they were physically together, as in an army
where soldiers could see or hear the next soldier in the column
ahead as they marched through the night to the battlefields of
Actium, Thermopylae, or Agincourt.

Maurice Bloch has given a vivid account of what clocklessness
meant in the Middle Ages—not that people were aware of what
they lacked, any more than we are aware of what we so decisively
do not lack.

> These men, subjected both externally and internally to so many
> ungovernable forces, lived in a world in which the passage of time
> escaped their grasp all the more because they were so ill-equipped
> to measure it. Water-clocks, which were costly and cumbersome,
> were very rare. Hour-glasses were little used. The inadequacy of
> sundials, especially under skies quickly clouded over, was notorious.
> This resulted in the use of curious devices. In his concern to regulate
> the course of a notably nomadic life, King Alfred had conceived the
> idea of carrying with him everywhere a supply of candles of equal
> length, which he had lit in turn, to mark the passing of the hours,
> but such concern for uniformity in the division of the day was
> exceptional in that age. Reckoning ordinarily—after the example of
> Antiquity—twelve hours of day and twelve of night, whatever the
> season, people of the highest education became used to seeing each
> of these fractions, taken one by one, grow and diminish incessantly,
> according to the annual revolution of the sun. This was to continue
> till the moment when—towards the beginning of the fourteenth
> century—counterpoise clocks brought with them at last, not only
> the mechanisation of the instrument, but, so to speak, of time itself.
>
> An anecdote related in a chronicle of Hainault illustrates admir-
> ably the sort of perpetual fluctuation of time in those days. At Mons
> a judicial duel is due to take place. Only one champion puts in an
> appearance—at none, at the ninth hour, which marks the end of
> the waiting period prescribed by custom, he requests that the failure
> of his adversary be placed on record. On the point of law, there is
> no doubt. But has the specified period really elapsed? The county
> judges deliberate, look at the sun, and question the clerics in whom
> the practice of the liturgy has induced a more exact knowledge of

the rhythm of the hours than their own, and by whose bells it is measured, more or less accurately, to the common benefit of men. Eventually the court pronounces firmly that the hour of 'none' is past. To us, accustomed to live with our eyes turning constantly to the clock, how remote from our civilisation seems this society in which a court of law could not ascertain the time of day without discussion and inquiry![8]

People could not be brought together without difficulty nor fine-plan and fine-trim their own time, because they did not ordinarily know what hour it was, let alone what minute. There was no advantage in trying to be punctual, because there was no way of knowing whether one was; in these circumstances there was no room for other values which in our age are so much taken for granted, such as punctuality itself, or the widespread credo about planning time so as to make the best use of it, or of regarding time as money even if it cannot be hoarded as easily. Such new values go with the irritable impatience about time which is the mark of the fully modern person. Patience is no longer a virtue, but a vice. No one wants to wait any more. Of the new spirit of Protestantism Max Weber said: "To waste time is thus the first and, in principle, the worst of all sins. The span of life is infinitely short and precious if one is to 'make sure of' one's election. To lose time through sociability, 'idle talk,' extravagance, even through taking more sleep than is necessary for health (six to at most eight hours) is considered worthy of total moral condemnation."[9] Before the clock it was easier to avoid being a sinner. Ignorance was a good excuse, if not before the ordinary bar at least in the higher court.

To get some sense of what life without timekeeping was like you can visit countries which are still, by our extraordinarily generous standards of provision, endowed with few clocks and those unreliable, and mark the exasperation of one's fellow visitors.[10] How thoughtless someone must be, they say, to be late by hours for an appointment, or not to keep it at all. The anthropologist Pierre Bourdieu gives the flavor of a thousand and one nonencounters when he speaks of the Kabyle peasant in Algeria as having "an attitude of submission and of nonchalant indifference to the passage of time which no one dreams of mastering,

using up, or saving . . . Haste is seen as a lack of decorum combined with diabolical ambition."[11] Here is another account, of an
agricultural attaché at a U.S. Embassy in Latin America:

> After what seemed to him a suitable period he let it be known that
> he would like to call on the minister who was his counterpart. For
> various reasons, the suggested time was not suitable; all sorts of
> cues came back to the effect that the time was not yet ripe to visit
> the minister. Our friend, however, persisted and forced an appoint-
> ment which was reluctantly granted. Arriving a little before the
> hour (the American respect pattern), he waited. The hour came and
> passed; five minutes—ten minutes—fifteen minutes. At this point he
> suggested to the secretary that perhaps the minister did not know
> he was waiting in the outer office. This gave him the feeling he had
> done something concrete and also helped to overcome the great
> anxiety that was stirring inside him. Twenty minutes—twenty-five
> minutes—thirty minutes—forty-five minutes (the insult period)! He
> jumped up and told the secretary that he had been "cooling his
> heels" in an outer office for forty-five minutes and he was "damned
> sick and tired" of this type of treatment. This message was relayed
> to the minister, who said, in effect, "Let him cool his heels." The
> attaché's stay in the country was not a happy one.[12]

The Parent of Other Machines

The first clocks were produced in response to an important time-
conscious constituency in the early Christian monasteries. Landes
has pointed out that by and large medieval society had no more
call for clocks than modern Kabyle society. Nature's rhythms were
all most people needed, except those devoted to Christian prayer.
In Judaism the devout are expected to pray thrice a day, but at
no set time. Islam demands five daily prayers, but they are at
times which can be judged without mechanical aid: at dawn or
just before, just after noon, before sunset, just after sunset, and
after dark. Christianity's more exacting schedule made clocks
highly desirable if not essential. When Saint Benedict established
his order, the Rule required eight offices of prayer: Matins, Lauds,
Prime, Tierce, Sext, None (at the hour of the death of Christ),
Vespers, and Compline. The hours for the different prayers were

marked by bells, which rang out through the cloisters and dor-
mitories day after day, whenever the bell-ringer knew what hour
it was. Punctuality was a moral imperative. Discipline had to be
strict in that first metronomic society if the power of prayer was
to be enhanced by everyone joining in.[13] The spirit was more
willing than the flesh, however: although all manner of devices
were used to help the monks keep to the Rule, until the clock all
of them were unsatisfactory.[14]

Once invented on behalf of the monks and the nuns, the clock
shaped other more general needs. Lewis Mumford in a famous
passage said:

> One is not straining the facts when one suggests that the monaster-
> ies—at one time there were 40,000 under the Benedictine rule—
> helped to give human enterprise the regular collective beat and
> rhythm of the machine; for the clock is not merely a means of
> keeping track of the hours, but of synchronising the actions of
> men . . . The clock, not the steam-engine, is the key-machine of the
> modern industrial age. For every phase of its development the clock
> is both the outstanding fact and the typical symbol of the machine:
> even today no other machine is so ubiquitous . . . The gain in me-
> chanical efficiency through co-ordination and through the closer
> articulation of the day's events cannot be overestimated: while this
> increase cannot be measured in mere horsepower, one has only to
> imagine its absence today to foresee the speedy disruption and
> eventual collapse of our entire society.[15]

Clocks ceased to be reserved for the mighty—in China, where
the first elaborate clocks were invented, time was thought to
belong to the Emperor—and they became cheap enough for more
general use. When people on retirement are given a timepiece, a
symbol of the time spent punctually and punctiliously in the
service of their boss, they are being given something which they
would never have earned without a clock. Clocks supplied an
answer to the old question about what time is: Time is what
clocks measure. This may be as circular as saying that intelligence
is what intelligence tests measure, but if people believe it about
time, as some most evidently do, it provides a framework within
which control can be exercised and some of the fear from Tagore's
night can be banished. If clocks can be adjusted by pointing the

well-named "hand" which extends far beyond the clock face, what is done within the intervals the instrument measures can be regulated too. The Babylonians, Julius Caesar, Exiguus, and Saint Benedict unwittingly prepared the way for the more modern spirits of James Watt, Benjamin Franklin, IBM, and the Pentagon.

The new servant of the sun has aspirations to become its master. A prosthetic device, yes: worn on the wrist, watches tie people to the celestial clockwork. Hundreds of millions of them, lying in what pockets, on what wrists, under what sleeves, revolve in time with it. Yet although they watch the sun, watches have established a kind of independence. They tick as remorselessly through the night as through the day, through the winter as through the summer, as though their environment is not going through vast changes in each 24 hours and each year. These perturbations are part of the lives of all other animals, but people have been encouraged by their clocks to ignore the daily and seasonal cycles and to pretend that there is nothing to contend with but metronome-paced linear time. We appear to have caged the sun inside a machine.

Acquired Habits

The customary set of attitudes has to be learned, sometimes painfully, by each new generation of children. These attitudes do not come from the womb but from society, in our case from a society which is more and more worldwide. Parents and teachers are the disciplinarians who by example and precept act as the agents of the new unified time. Jean Piaget's studies of young children have shown how much intellectual work has to be done by them in order to understand simultaneity (that different events happen at the same time), then succession (that one event happens after another), and then duration (how long a series of events takes). Children also have to be taught to think backward in the same linear fashion that they think forward. Piaget says (rather misleadingly), "it is only the contents of time, i.e. the external or psychological events, that are irreversible—time itself is a system of reversible operations."[16] By "time itself" he is referring to the sort of time which can be recalled in memory, as "time" is put into reverse. But about the effort needed there can be no question.

What has to be learned is not arbitrary, of course. In all cultures, natural cycles are the main armature of time measurement, which is made up of moving parts like a human body. Our metronomic culture, however, almost totally conceals what it has been built upon. How lost we would feel without the consensus, how dizzy if hours and minutes were abolished and new notations put into their place based on new tempi! No government in the world would be strong enough to impose that.

Such changes are not likely to be made, nor even proposed. The only recent reforms that have been successful have been those that brought older or other systems of measurement into line with the Christian, or made the moon give way to the sun. The old-style Chinese calendar was made up of 12 lunar months of 29 or 30 days each. An intercalary month had to be added every two or three years to make good the annual deficiency. The extra month was not added on to the end of the year but inserted between two other months so that there were two fifth or sixth months. There was a reform in 104 B.C. but after that none until 1912, when the new Chinese Republic discarded the whole system in favor of the same Gregorian calendar that England adopted in 1752.[17]

Modern supporters of temporal homogenization, for what seems like the most defensible rationality, have so far held off all attempts to break with the Christian reckoning. The new decimal system introduced in the French Revolution was certainly far-reaching. From 1793 to 1805 France enjoyed (or suffered) a 30-day month divided into 10-day weekly cycles called decades, a day divided into 10 hours, an hour of 100 minutes, and a minute of 100 seconds. The reform was unpopular and was eventually abandoned, even though the new names for the months devised by a poet hired by Robespierre deserved a better fate than the new names for months thought up by the Nazis a century and a half later. Stalin's introduction of first a five-day and then a six-day week in 1929 and 1931 was likewise discarded. So were the many proposals of reformers from Auguste Comte onward who wanted a new, and on the face of it, sensible calendar, with a year of 364 days divided into exactly 13 months of 28 days each, and each month into 4 weeks. One annual blank day, say Christmas day, would have to be excluded, and a leap day added in leap years. Comte, often regarded as the founder of sociology, hopefully gave the annual blank day the name of the Festival of All the Dead and the four-yearly leap day the name of the Festival of the Holy Women. The only successful reformers have been the small-timers, like the proponents of Daylight Saving Time; and even they have had to overcome immense opposition. People defend their far-reaching system of time discipline in industrial societies because they have become accustomed to it, even though it puts them under continuous and scrupulous scrutiny, from the time they clock into the records at birth to the time they clock out at death.

As it is, by a process of trial and error which has to be gone through by each generation of children, we have built up over several millennia a comprehensive counting scheme much more universal than the new world language. The common scheme extends down to the borders of the infinitely small within the day and up to the borders of the infinitely large within cosmology. It is all thoroughly linear. The line is not only inside people's minds but seemingly also outside, dissociated from human events but ready to assign them a proper place within its coordinates. In the fourteenth century clocks had only an hour hand, so that when punctuality mattered a start had to be made on the hour rather than in between; then the minute hand was added, then a second hand. What next? The precise second in any today can already be determined by the clock, the numbering of today by the calendar, and in any today the age of oneself and everyone else and indeed everything else as well by the general reckoner. The universe, and everything in it, has been enclosed in tick-tocks.

Contemporization of Time

Imagining time on a spatial linear scale, which has become so habitual to many people that it no longer seems imaginary, has been greatly enhanced by its extension forward and backward. There has been a massive "contemporization" of both the past and the future into a present which is being continually extended. Communication has improved not only across space but across time too, between the present and the past and the present and the future. The past has grown ever faster through the technology which has created the modern kind of collective memory. In simple societies, without writing, the collective memory did not extend back much further than individual memories. Such societies could not preserve in detail the works of an elite going back thousands of years: no Aristotles, Julius Caesars, Saint Augustines. Their heroes had to be spoken for, rather than being allowed to speak for themselves. The number of generations that could be recalled in the oral tradition, back to the origin of the tribe, had to be kept more or less constant by dropping a generation for each new generation that was added.[18] The past had to be truncated in the same way, even when it was not packaged up into generations; to some extent that happens even with us, as that segment of the recorded past which is not made more vivid by the oral accounts of parents and grandparents and great-grandparents and their contemporaries becomes that much more distant. Interest in the eighteenth century, as it has receded into the past, has slipped away, even among scholars, and the nineteenth is suffering the same fate. The previous century remains a golden age of certainty and hope only briefly. Because every person alive is a candidate for nostalgia, perhaps even the twentieth century will confound today's pessimists when it takes its place as the previous century, bathed in another glow, before it too disappears into the dust of the decimalized column; its obscurity will be much less than it would have been without the exponential stacking up of millions of memorial tablets.

The invention and spread of writing has made it more difficult for memory to perform its necessary erasing function of keeping the present fresh by not remembering too much of the past. Writing has become a potent instrument of social control without which no modern state could act the tyrant, or even the shepherd. When writing all had to be done by hand, the audience for the scribes' handiwork and illuminations was very limited. Some but not many could have access to some but not much of the past. But when printing followed the example of human habits and learned to produce replicas easily, more could be preserved, and each book printed could act as the parent for others in a book population expanding in geometric progression. Thereafter, attitudes were never the same. When the Bible was printed, in vernacular translation at the command of Protestant rulers who made what had been a cosmopolitan resource into an instrument of nationhood, the Bible also became the "treasure of the humble."[19] It gave spiritual dignity to householders who could own and read a copy, and encouraged a hundred different interpretations where before the church had, with some success, maintained a monopoly on the truth. Sects multiplied; existing ones became stronger and printed their own commentaries on a scripture which was not so unchallengeable as it had been. Likewise, printing brought to larger minorities, if not to the masses, copies of the classics, such as Ptolemy's *Great Composition,* the ever popular maps of the known world, the new works of science, and contributions to journals such as the Royal Society's *Transactions* with their pictures, diagrams, and figures.

Each book reproduced lays claim to a niche in the store, which therefore grows and grows, tempting the present to try to immortalize itself, always aided by the earnest efforts of journeymen whose existence is dedicated to the piling of the past upon the past. Historians are the busiest, ever finding new nuggets of fact and opinion which others before them had passed over and pushing back the frontiers of the past by giving voice to people who were not previously heard. Michael Ventris did this on the grand scale when he deciphered the Linear B tablets from the thirteenth and fifteenth centuries B.C. Archaeologists dig into the past more literally, and tribes of scientists have been doing their bit with exemplary diligence. In the seventeenth century the Archbishop of Armagh announced that the world began exactly on 23 October 4004 B.C., according to his scholarly interpretation of the Bible. Two centuries later physicists had put the date at more like fifty million years ago. By now, the almighty registrars have pushed back the birthday for the human race and allotted the common platform a birthdate four-and-a-half billion years ago.

The journeymen are not without doubts about their trade. The historian of printing Elizabeth Eisenstein has written that "the voracious appetite of Chronos was feared in the past. A monstrous capacity to disgorge poses more of a threat at present."[20] Another says, "Previous generations *knew* much less about the past than we do, but perhaps *felt* a much greater sense of identity and continuity with it."[21] According to Maurice Halbwachs, "history indeed resembles a crowded cemetery, where room must constantly be made for new tombstones."[22] Such doubts about the public prints are not kept private.

Contemplating the more distant past has become a growing scholarly endeavor, contemplating the more immediate past a growing industry. The industry's product is the past not just of any locality but of the world. As world population expands there are ever more people to have news about. The billionfold past is carried into the present by means of newspapers which present each morning such a generous account of the hours just gone by that if all of it were read conscientiously nothing would be done today except keeping up with yesterday. The industry even supports scholarship, by converting trivia into vintage. A few copies of the newspapers are frozen in archives so that they can be dragged into some future present in 10 years, 100 years, 1000 years, 10,000 years' time, not by other journalists but by eager scholars making the commonplace extraordinary. Fortunately, only a few copies are treated like that. If all the newspapers ever produced were stored in their original form there would before long be room for nothing else on the planet except the historical record. Cities would be buried by the aspirations to immortality of the past. Computers and microfilm may have arrived just in time.

Radio bulletins and television newscasts on the hour and the half hour and the quarter of an hour, like those on the BBC World Service, tell their beads so incessantly that they threaten to make Mumford's old-fashioned clocks quite obsolete. They keep playing back the past irrespective of whether anything worth noting has happened in it. The movements of certain key actors are followed almost minute by minute, not because they are doing anything remarkable but because they have a keen sense of the sequence of deadlines the media must meet throughout the day. These deadlines are signals for the springing up of yet another pseudolife given to someone whose words and smiles are transfixed in this brilliant mummification (or telefication) of the past. Communication with the dead is being improved almost as fast as it is with the living, and with ever more of them as written words multiply, pictures bubble, and other forms of recording burgeon. Sometimes their voices retain their eloquence. Shakespeare can no longer talk to his Dark Lady, but he can talk to ours. It has been said of Rousseau, influential though he was in his lifetime, that "with respect to ideas properly so called, almost all those of Rousseau have germinated after him; they have blossomed about us."[23] If true of Rousseau, is it not true of many others as well? Does Ronald Reagan carry more weight than George Washington? Mr. Gorbachev than Marx? Any Poet Laureat than Homer? Anyone than Shakespeare? If one of the prime tests of life is that other people can vouch for it, Washington, Marx, Homer, or Shakespeare are in some people's minds much more alive than the living. The dead can also sometimes be more satisfying companions, offering a stability, firmness, and constancy that no living person can aspire to.[24]

The immortals may seem to belong to a different spiritual universe from the rest of us. They have a certain godlike quality even if, as with the marvelous painter of the black bull in the Lascaux caves, we do not know more of him or her than that they are our intellectual ancestors; many of us can indulge in ancestor worship all the more unrestrainedly because no one can accuse us of doing so to curry favor with them. We cannot be lickspittles of the past, however much some of us would be glad to be. If we are egalitarians we may condemn the past grandees for belonging to a narrow elite instead of the common democracy of the unknown. But revolutionary sentiments will not change the past. We cannot open up the arena of the present to the countless millions of the unknown whose bones lie not so nobly under the waters off Cape St. Vincent in the fields of the Caudine Forks, or in the plague pits of London. It is too late to bridge the gulf between the common man and the illustrious—between Nelson and the lowest of his seamen, between Hannibal and the meanest of his elephant riders, between Defoe who described them and the unknown Londoners who were thrown into the plague pits.

Can any one-way communication really be considered immortality? Perhaps not. But to judge the dead by their characters when alive, some of them might, if we could consult them even more easily, express themselves as entirely happy with the one-way communication, and perhaps acknowledge they were right to seek the fame which cannot be killed. They are in the same situation as TV personalities, who are dead according to one criterion because they have voices without ears, but are apparently alive all the same and enjoying sending into the air a one-way stream of vivacious chatter. Their tongues at least are immortal, if not their ears. TV is the realm of the Living Dead, the best proof of the arbitrariness of the distinction between life and death. There is no way of knowing whether the persons you watch are live or on tape, or whether they could take part in a phone-in program. Anyway, no one cares how much dead people repeat themselves on the screen. We would be mightily alarmed only if they did not repeat themselves exactly. Imagine tonight's audience for a Hitler if he were guaranteed to change on the TV a speech he had made before at the last Nuremberg Rally. Only the historians would refuse to tune in.

Many of the famous from earlier eras would have been TV personalities too. While they were alive they were not humble enough to show much interest in the opinions of others, nor to suffer foolish arguments gladly. Good speakers, they were not good listeners. They could contemptuously regard us as just survival machines for carrying around their immortal ideas. For most of them television would be the ideal situation, where there is not the remotest danger of being interrupted and where they are on good terms with us without giving a thought to whether we are on good terms with them. For the minority who are not vain and who wriggle in fame, not having sought it or having had to put up with too much of it, it might seem like Hell, an everlasting life with no possibility of escape into the peace of oblivion.

There is no longer any now for any of them, except of course the now we so generously lend them. Unless they can communicate with each other, they are encased only in our memories, individual and collective, so that the most humble soldier of today, reversing the social order, can hold Alexander and Napoleon prisoner. They speak to us from our past where they belong, and cannot surely have our sense of the future. It is only we, the still living, who can ride the crest of the wave which is bearing the present into the future. This is the moving terminus at one end of the line, where we exactly are; whether asleep or awake, old or young, we all are suspended on the same crest with our five billion fellow riders. Though fortunate that we can travel with so much of our collective history to guide us, none of us can bear our immortal memories faster into the future than we can travel ourselves. If the speed of light is the absolute limit for all velocities in space, the speed at which time is unfolding is the absolute limit for the speed at which we are all moving.

Overloading Time

Gregory the Great praised statues as "the books of the illiterate."[25] Even if he could have praised the media on the same grounds, he would surely have been worried by the sheer scale on which they preserve the past, and especially the "stars" from the immediate as well as the less immediate past. Back in 1965, outside the workplace the average American spent nearly 6-½ hours a day "consuming" information from the mass media—or, should one say, being consumed by it? Twelve years later, in 1977, the information load was up to 8 hours. Even with so much time given to it, the daily task was a formidable one. The number of words from which to choose, coming from radio, TV, cable, records and tapes, movies, newspapers, magazines, and books was eleven million per day in 1980. Of these, even when concentrating on the task, the obedient citizen could only read or listen to forty-eight thousand.[26] More books have been published since World War II than in all previous centuries put together. The explosion of knowledge mirrors the cosmos.

In addition to input flooding in from the main media, an almost equally generous input from the same technology in a more individualized form has continually increased with the size of people's social networks. The expansion of the common time zones by transport and telecommunications has steadily widened people's range.[27] They are no longer confined to their immediate neighborhood. Connections are maintained over ever larger territories by messages for each of us, carried from others who are or would like to be in our networks by mail, telex, facsimile, and above all by telephone. Monasteries were governed by the sound of bells booming from their belfries to call people to prayer; modern homes are governed by the sound of the bells of the telephone, summoning people to attend, not for set hours at Compline and None but at any hour whatsoever. It is one of the triumphs of the culture that people are so fully habituated to this particular machine, second only to the clock in its penetrative power. Since telephones stopped being used mainly as a substitute for the telegram in emergencies, they have reduced isolation in rural areas, acted as baby-sitters, increased peer group contact, and allowed members of extended families to be extended in space without losing contact with each other. They have reduced isolation to such an extent that more and more, people cannot be on their own at all. "At your beck and call" has a new meaning. For the Nuer, in Africa, according to Edward Evans-Pritchard, "the daily timepiece is the cattle clock, the round of pastoral tasks, and the time of day and the passage of time through a day are to a Nuer primarily the succession of these tasks and their relation to one another."[28] Our clocks are more insistent.

People in industrial societies cannot shelter themselves from their daily timepiece even behind walls of answering machines. The growth of communications is so rapid that it is difficult to imagine what will happen in the next century, unless computers, which have generated so much information, come to our aid by dealing with it on our behalf, thus forming a new self-sustaining system which hardly needs any human monitoring at all. Millions of computers could then get the habit of talking to each other incessantly, especially in their private computer time at special bargain periods during the night when the nominal masters and mistresses are safely asleep and no longer able to overhear what their slaves are saying. Computers could become the citizens of night society.

When I was visiting Massachusetts, I received a telephone call which at first raised my hopes. "Good morning," said the girl's voice. "Good morning," said I, wondering who the girl could be.

"I am making a survey on the ways people use toothpaste in America. The first question I'd like to ask is, what kind of toothpaste do you use most?" I realized, with a rush of disappointment, that I was being talked to by a computer and banged down the receiver, deciding to keep my toothpaste to myself. Clearly, one advance has run ahead of its complement, as so often happens with technology: another telecomputer is needed that will silently answer such calls for me in the middle of the night, and which can really answer, not just record the questions to plague me with when I wake up. The thoroughly domesticated computers of the future will have each other's numbers. Mine will then be able to ask the ad agency's computer what sort of toothpaste *she* uses, at least if both partners to the dialogue have the right level of artificial intelligence and can graciously simulate sympathy for each other's teeth, even if they have none.

Without such relief, there is not much reassurance about how to stop the present becoming ever more crowded with the past, let alone with the future. The ever increasing number of people in the world and their ever more diligently recorded interactions could leave no time over after the maelstrom of information has been dealt with. The mass media are machines for crowding the present with the near past and universities are machines for crowding the present with the far past; but they will both be doomed unless ears can be invented to match their mouths.

The measuring instruments have been increasingly used to the opposite effect. The future is not allowed to plod gently along into the present at a decorous pace; it is being sucked into it. For the future is in one vital respect no longer the unknown. It has been strung out into an infinitely extensible line of future presents by being given precisely delineated notations in advance. No bit of the future can escape the grasp of the tempographers; they have measured out, before it has happened, the only feature of our environment which can be made to appear almost wholly regular. The great world-spanning agreement about time measurement has accorded the future an axiomatic predictability. People can be required to make the most elaborate plans for days, weeks, months, years, even generations ahead. It is done in every organization and by all states. In the United States a president is barely inaugurated before the cycle leading to the next election, day by dread day, begins to revolve—although such predictability is admittedly almost unique among political systems.

I imagine no one knows who will be struggling to become President of the United States in the autumn of 1996. But we all firmly expect that there will be an election day that November, and a 7:05 train that morning and a 5:50 that afternoon for people who want to take their cocktails early. If you are an obsessive planner, have a good doctor to keep you going, and are not expecting to be too partisan, you can invite a friend right now to join you and your doctor on that day at five to six, by which time the ice, left for five minutes, will have melted enough to slip obligingly from its tray. All can be figured out in advance. The ice is ready now. Appointment diaries for 2096 could be printed immediately with their weeks numbered 1 to 52, if anyone who thought inflation was going to balloon printing costs had enough capital and storage space for them.

Some professional people can anticipate not only where they should be in their careers decades hence but, over shorter periods, settle what they should be doing even more precisely. Here is an account of an American doctor who practices in and out of hospitals:

> Barry's life is scheduled far in advance. He knows when he will take his vacation next year, which weekends and nights he is "on call," which mornings he will make rounds in which hospital, and the hourly contours of each day . . . The schedule provides slots for work whose size is limited in several dimensions. Certain problems can be dealt with in the hospital, but not in the office, or in the afternoon hospital working visits, but not in hospital rounds, or in the intensive care unit, but not "on the floor," and so on. Barry tries to fit the work into his schedule. If parts do not fit, they are abandoned or postponed, or the schedule expands, extending Barry's day, deferring and reordering other appointments. He "runs behind." Barry is oppressed by two sorts of temporal constraints: too much to do in the allotted time, and the urgency of action imposed by what are too often literal "deadlines." Time is differently shaped accordingly: excess draws it out, imminence intensifies each moment.[29]

The business executives I have known are different. Their time is also constrained. They need to keep rivers of information from the past flowing into their present in order to control their future and that of others. They also are often "running behind." But they do not know in the same way what they will be doing at a given point in the future. They may follow some rudimentary cycles for higher priorities—board meetings on the third Monday in the month, internal meetings on Tuesdays, golf on Wednesdays, Thursdays, and Fridays—but otherwise special arrangements have to be made for every day of their lives. They are at the opposite extreme from people in a small English village who do not have to make any arrangements in advance to meet each other: they can go to the pub when they know someone they are after will be there or look over the garden wall when he will be digging his potatoes or knock on the door knowing she will be in the kitchen. In such a place behavior may be so much habitual that people can predict with assurance others' whereabouts and activities and use the knowledge to get hold of them when they want to. They have more time in the present partly because so little of it has to be spent forecasting and negotiating the future.

The executives have no such assurance. Almost every prospective hour has to be organized, the task being a great consumer of time, especially for the time brokers, secretaries and assistants whose job is to guard the appointment diaries and buy, sell, or exchange the future time of their bosses. For portions of the future, full-scale diplomatic negotiations have to be opened up, so that a slot two months ahead can be found which suits, say, five busy people. No negotiations are ever closed until the clock has rat-a-tat-tatted its way into the present. All treaties about the slot are provisional, dependent on fulfillment of the expectations harbored when the first treaty was made. When priorities change the treaties are revised: long-standing appointments have to be scrapped and painstakingly remade. The future does its tantalizing best to escape from its putative straitjacket, sometimes succeeding so well that it seems to be forcing the present backwards. If people spend more than a day just planning a day that is to come, they can get caught in a regress from which there is no escape. It is rather like the problem faced by Tristram Shandy in Sterne's novel, which found that because it took him two years to write the story of the first two days of his life, the material for his biography would always grow at a faster rate than he could record it, and he would never be able to complete it.[30] If each day in the future takes longer than a day to plan for, future planning will never come to an end and more and more of the present will be squeezed away, with the consequence that people will lose time steadily.

Such negotiations have produced a system of status measurement as time assumes a more and more central place in modern society. Some people very high up in the hierarchy are "one-dayers": they are so important that a request for an appointment is granted almost immediately, within the day, even though lesser mortals—say, one-weekers who until then thought they were safe—have to be shuffled out of the way with as much grace as possible to make room for their time-superiors. Subordinates hardly rate: by definition, their time can be commanded rather than negotiated. Lower-status people are made to feel they hardly have a right to inhabit the same present as their superiors. The difficulties arise when people are of equal status, or near enough, so that they have to use a lot of energy in overt dominance displays of how important they are compared to the other person.

In a small study I made of senior executives in Britain it was clear that many of them were under extreme time pressure. Executives were likely to overstretch themselves by committing themselves to an 8-day week or a 30-hour day, always trying to do too much. When their elaborately planned futures arrived they were rushing from one event to another without time to savor anything or take a well-considered decision; they did not allow themselves time to think about the future of their companies, their colleagues, or themselves; they were continually harassed, even though they were never supposed to look as though they were; and they had no holidays from the pressure of time. Time had to be rationed every day in the calendar, and when any of them said, "I cannot do this or that, I have no time," he meant it even though it was nonsensical. He had, or rather should have, the same amount of time as anyone else, instead of being permanently out of breath as he jogged past the stations in his calendar. Claiming to have no time was a means of asserting status, measured by panting.

People always have time to tell you they have no time, like those who go to great lengths to show how valuable their time is by telling you they travel on the Concorde and letting you know they have no time for economy-class travelers like you. They are always running behind time or running out of time, and all is as it should be as long as they keep running. All is well as long as you are not seen waiting for anyone or anything. If you go against the rule and say, "Come in whenever you like—I'm very free," others are likely to accuse you, under their breath, of false humility.

The Time Famine

Given the prevailing linearity of our society, what I have been saying about time pressure on executives at work can be generalized, especially to older people with larger networks and more demands upon them. Economists who have followed the lead of Lionel Robbins and Roy Harrod in England, of Staffan Linder in Sweden, and of Gary Becker in the United States have found that the axiomatic and automatic consequence of a rising standard of living is that time becomes scarcer. As productivity rises, abstention from work costs more than it used to, the cost of the leisure being what could have been earned if the time had been spent at work instead. This cost is, in economists' terms, the opportunity cost. The near-catastrophic consequence is that time is getting increasingly more scarce, and will do so as long as productivity goes on rising. Although people are living longer, their lives may be getting steadily shorter in subjective terms. In a time famine it is not difficult to feel the pangs of the new hunger. Eventually, perhaps time will become so scarce it will go off the market entirely.[31]

One would expect this relationship between time and riches to have some effect upon the hours of work. The effect should depend upon which is stronger, the income effect which provides that as incomes rise people will be able to achieve any given standard of life with less work; or the substitution effect which operates so that the more they work, the higher the standard of living they will get and the more goods to substitute for leisure.

According to economists, because working hours have fallen for the majority of people in industrial societies, the income effect must have swamped the substitution effect. In industrial societies people have at least had more leisure, even if it is more expensive in goods and services forgone, and even if they are more and more occupied with the unpaid work which a higher standard of living generates. It takes so much work to maintain their property in houses, cars, hi-fis, TVs, videos, computers, microwaves, books, mixers, freezers that they may sink under the burden of civilized living. "If I keep a cow," Emerson wrote, "that cow milks me."

For some there is the double burden of this harassment in leisure and no reduction in hours of paid work. A few doctors, lawyers, and other professionals have acted according to the economists' "backward-bending supply curves" of labor—that is, they have worked fewer hours as their incomes rose; but not all those who can choose how long to work have reacted this way. Those who work in their own time are apt to work in all of it. Other professionals cannot choose. Managers and others are bound to organizations which set their hours for them. Such people may have lost leisure rather than gained it. Whatever the other advantages of their lives, they are the victims of time; they are harassed at work and away from it, even though they are able to spend a great deal of money on the fruitless task of trying to "save" time.

I see nothing wrong with the economists' position except that their reasoning base is too narrow, tied too firmly to the standard of life. The outlook is in fact even more grim than they allow. If the cost of doing anything is what you give up in alternatives, then it follows that anything which you contemplate as an alternative adds to the prospective cost, or as I have been putting it, the pressure on time. Every choice made is a possibility, or a whole raft of possibilities, destroyed. The contemporization of the past and the future on a worldwide scale through the miracles of modern communication has made people more aware of the many alternative ways to spend any hour or day, and so more conscious of the shortage of time.[32] They are more aware that in attending to one bit of information they are denying themselves the possibility of attending, with full concentration, to another. They are more aware of other places in the world where they might be, with other men, or other women, at other meetings, at other conferences, at other exhibitions, on other trails, reading other books, on other moonlit nights. They are made continuously aware of what they haven't got, which makes riches such a trap.

The worst of it is that many of the tactics used to deal with the fundamental shortage of time seem to be self-defeating. A common device is to do more than one thing at once. People distract themselves by watching a horse race on the TV while they knit and shout at the children, or plug into the music on their Walkman while they walk to the shops with a glazed, faraway look in their eyes. After dinner a man "may find himself drinking Brazilian coffee, smoking a Dutch cigar, sipping a French cognac, reading the *New York Times*, listening to a Brandenburg Concerto and entertaining his Swedish wife," and so further aggravating the problem of his Swedish wife.[33] Another man may so thoroughly confuse time with space that he thinks he is slowing up time (or saving it) by traveling faster than other frantic motorists also tilting at the traffic lights.

The symptoms of the famine are all around us. In my home city, London, I have only to telephone an executive, a senior civil servant, or still worse, a professor—they are all busynessmen and women now—to be told that this next week or month is quite impossible, there is an important meeting in Glasgow or Brighton or Washington that has to be prepared for or gone to—as if one needs to be humiliated cut by cut by a tally of all the other calls on his or her time which are so much more important than mine. Or I'm told that if I can phone back at the end of May, say on the twenty-eighth, there does seem to be a window at 11 o'clock. If I put up with the put-down and succeed in getting into the sanctum, I feel I ought to get up and go as soon as I sit down, because the man in the office is making such frantic efforts to show me how valuable his time is by talking on the telephone, scribbling notes, and reading memoranda and signaling wordlessly to me that if only I will wait a few more moments a window *may* open up before the next phone call tinkles into the room, and through that window the two of us, so privileged to be alive together, may be able to exchange a word before the exploding universe tears us apart again. His left shoulder is plaintively raised to show that it has spent thousands of hours clamping a telephone to his ear in order to free both hands of this orthograde man for simultaneous writing, paper shuffling, and calculator tapping. If I fail to do more than exchange increasingly pained glances, I may be able to get him later on his car phone, whose novelty I can hope has kept him from installing an answering machine or an answering secretary as a fixture in his car. While he chats to me he will be able to listen to the stock-market prices on his car radio and mouth to colleagues through the window who are trying to catch him as he drives out of the building that he's too busy just at the moment, but there may be a window later on. He is hell-bent on making time's winged chariot fly ever faster.[34]

Divorce from Natural Rhythms

The increasing control bestowed by the clock and the calendar has gone hand in hand with a campaign for a new independence for humanity—a hundred years' war against nature. Years have been divided into equal days, days into equal hours, hours into equal minutes, minutes into equal seconds, seconds into equal nanoseconds; these are the beats of our metronomic society. We can rely on them all the more because we have cut (or at any rate cut down) our ties with the natural rhythms which used to mark time for our forebears. The natural fluctuations in light and temperature have been damped down as our dependence on mechanical clocks has grown. If we were not able to buffer ourselves against these cycles, we would not be able to give such primacy to the clock or feed so vigorously the vanity that our species is on its way toward mastery over the environment. Much of the GNP of every industrial country is therefore siphoned off for the great cause of divorcing ourselves from nature. Poor countries cannot embark on this expensive mission: they do not put so much of their resources into keeping cool or warm and, above all, lit up. Rich countries on the other hand can benefit from the transfer of energy from great holes dug in the earth's crust in one part of the world to challenge the dominion of the sun in another part. The magic lattice of New York's night stands on Arabian sands. Nonrenewable fossil fuels—oil, gas, coal—which store the sun's energy are piped from the primeval past into the pulsing present to maintain people in contemporary nights and contemporary winters in the temperature-conditioned urban fortresses to which God did not call them. It is now only young children He sends to bed at dusk.

So fundamental has this resistance to natural rhythms become, even if it is artificial, that it can be called a sociostatic process. I have said that the homeostatic does not supply as good a construct for bodily processes as the homeorhythmic. But when people are habituated to counterfluctuations, the notion of sociostasis will do well enough. Cyclical fluctuations seem to be such an affront that people take it for granted that a great deal of effort should be put into maintaining the environment, at any rate their built environment, at a more or less constant temperature and luminosity for as long as their eyes are open. Evening out natural fluctuations has become an egalitarian enterprise which it is heresy to question. People will not wear more clothes in the winter to keep warm if they have the option of heating much larger spaces in which to sit with almost nothing on.

The distaste for night is more understandable, especially in far northern and far southern latitudes where the natural fluctuations in the length and temperature of day and night from one season to another are so much more extreme. The industrial revolution had to start in such latitudes: the gains were so tangible. It could not have started in Southern California, though it has evidently extended there. The tendency to sociostasis has spread from the darker to the lighter and from the colder to the hotter places; and people have undertaken one of the great migrations of the twentieth century, toward the equator, or at least nearer to the Mexican border or the Mediterranean.[35] If people are no longer too cold in winter, they have the saving grace of being too hot in summer, and so can quite properly spend the resources they do not need for radiators in the winter on air conditioning in the summer. Conditioning is the right word: the air has been forced into the habit of temperance.

Such is the time lag in adjustment to new circumstances that people still behave as though climate control had not yet been accomplished. The tourist economies of the world are blessed by people who keep themselves warm or cool to a perfectly satisfactory level in their own houses, cars, offices, and factories all the year round but still like to travel for a vacation nearer the equator where for a few hours they can frisk around outside and be not much colder or hotter than they are inside at home. They can spend the rest of their time being splendidly borne around in air-conditioned buses or sitting in their hotels planning their next trip, while being kept continuously at the same beneficent temperature and being fed the same year-round strawberries that have become their right. All over the world, lines of people on holiday get into their constant-temperature cocoons at eight in the morning and ride through the traffic at the same hour as they ride to their customary destination at work. They are witnesses to man's solicitude for man in one of the respects that seems to matter most, temperature; that is, apart from the old and poor who are allowed to die of hypothermia.

The Incessant Species

For the greater part of our history people have had no alternative but to put up with night. Darkness was an act of God, to be accepted by diurnal animals, including us. No one would have predicted that humans would become the first animals to try to be both diurnal and nocturnal—the incessant species—at a time when only rich or otherwise powerful people could carry candles everywhere they went. A turning point was when Pall Mall in London became the first street in the world to be lighted by gas, in 1820. The gathering in of night has been more and more contested since, as fire for light has been urbanized and domesticated, first in the form of gas mantles and then in the form of the triumphant electric lamp. Artificial lighting has transformed people's lives. It has created a sector of the day, and a sector of society, devoted to leisure: people's labor time is sold to others in the day but kept for themselves in the evening.

It has also created a new society, the society of night. People can go out and stay out at night, in lighted buildings or in floodlit stadia, and indeed see more people's faces under artificial light than they do by sunlight. They can listen to radio or watch videos all night long. In defiance of circadian propensities many jobs have been switched out of ordinary daylight hours into the night. When, in 1867, Marx pronounced night work a new method of exploiting labor, the practice was in its infancy.[36] Murray Melbin of Boston University is one of the few sociologists of the full-grown night society. He has summed up the incessant organization that America has provided for the incessant species:

> Today more people than ever are active outside their homes at all hours, engaged in all sorts of activities. There are all-night super-markets, bowling alleys, department stores, restaurants, cinemas, auto repair shops, taxi services, bus and airline terminals, radio and television broadcasting, rent-a-car agencies, gasoline stations. There are continuous-process refining plants, and three-shift factories, post offices, newspaper offices, hotels and hospitals. There is unremitting provision of some utilities—electric supply, staffed turnpike toll booths, police patrolling and telephone service. There are many emergency and repair services on call: fire fighters, auto towing, locksmiths, suppliers of clean diapers, ambulances, bail bondsmen, insect exterminators, television repairers, plate glass installers, and funeral homes.[37]

He has called the switch in timing "the colonization of the night," in which the empire of the day has been extended into the night. He talks of night as the temporal frontier which is crossed by the same kind of tough and independent people who crossed the spatial frontier of America in the nineteenth century. He says that California's main cities show great activity in the dark hours, as if the flow across the continent swerved into the nighttime rather than spilled into the sea. When I was in Los Angeles I tried to cross the same frontier myself.

I ask at the police department if I can spend the night with a police officer. I am rather surprised to find I am assigned to the vice unit. Perhaps they think that a member of the House of Lords would be especially knowledgeable about that. Sergeant Hofstadter is my companion. He takes me out in his patrol car, a Cougar that is unmarked so that, on his vital mission, the prostitutes he is chasing cannot duck into side alleys when they see danger. A lean, handsome man in plain clothes with two degrees in public administration and now a highly trained member of the vice unit, he is locked into a standard clock-bound bureaucracy. He signs on at 6:30 p.m. and stays at it till 3:30 a.m. Every half hour of his time is logged. His car radio crackles incessantly with messages from headquarters and from other police cars. "White girl standing on her own on the corner of Thirteenth and Wallace. Saw her talking to a man in a blue station wagon. Worth investigating." He answers, "This is Hofstadter, Vice, Pico and Figueroa, I'm off." He accelerates, dropping down to walking pace as we approach the girl and see her get into the large blue car with two men. When the car turns into an alley and parks, he stops, gets out, pops his head around the corner and stands there, waiting patiently until the car begins to rock up and down. At that squeaky signal he walks up the alley, shines his flashlight onto the girl and one of the men in the back of the car and arrests them both, at 9:27. The girl has sixty dollars tucked into her white boots, which fall out as she gets dressed.

Several arrests for "lewd conduct" before 10 o'clock would have been routine for him on other nights, but on this night in the lunar month an arrest is unusual. The moon is nearly full and the night sky is cloudless. He told me before we went out that I had chosen the wrong night: there would probably be little activity in the streets. Clients are put off by the moonlight and so are the prostitutes. In the backs of cars or in the dark alleys where they sometimes lie they are more likely to be seen, and, if they do risk it, they are more likely to be arrested. So full moon is nearly always a slack time for "vice," and winter also quieter than summer, except on unusually warm nights. At full moon he does not ask young women police officers to dress like prostitutes and walk the streets to entrap men: the moon's effect would be just the opposite of the one it is supposed to have. Rabindranath Tagore might not have approved of this particular sensitivity to the heavenly body, but perhaps his spirit is with us, and with the women. They are responding to the moon not for romance but more as fishermen have done since time immemorial, not venturing out at full moon when they would be more likely to be seen, by the fish in the one case, by Sergeant Hofstadter in the other.

Melbin noted that there was more helpfulness and friendliness on the old frontier, quoting an English traveler who said, "Even the rough western men, the hardy sons of the Indian frontier, accustomed from boyhood to fighting for existence, were hospitable and generous to a degree hard to find in more civilized life."[38] With this in mind, in the middle of another winter's night I went to Venice Pier, which juts out into the Pacific from the same city.

There, eight amateur fishermen and fisherwomen are standing around an oil barrel which is being used as a brazier, burning paper taken out of litter bins, while others standing by the rail at the water's edge watch their rods for them. They hold out their hands to the warmth. The light from the flames flickers over their faces, especially welcome in the darkness. They talk, rather slowly, to each other and to me. One black man says the pier is like the place where he was brought up in Mississippi, without the boon of electricity. His family was the first to own a television set, although it could not be used. Amazed people came to stand in the street and look long at the strange object through the windows of his house. His father shone a lamp on it for the onlookers to see better. A black woman whose eyes flash in the firelight says that none of them has ever been attacked on the pier. They feel quite safe. They would all help each other, if needs be. She and her woman friend spend every night there. They know the names of the men who arrive in a group around 4 a.m. with their larger fishing rods and their thermoses of hot coffee. They are the shark men. Later, at 5:30, come the bonito men. The ruling class may have gone to bed; the fish, and the people who wait on them, are going through their customary rounds.

The examples I found were not people who had made night into day but those who still acknowledged the special quality of night. Their sky was still the night sky.[39] Other people awake through the night inside lighted buildings, in factories, power stations, telephone exchanges, wholesale depots, hospitals, would be more typical; even if they are still a minority in the whole population, they are the larger minority.

The majority are also more buffered against natural cycles than their ancestors. They can look out of the windows of their apartments or offices—factories are often without windows and use artificial lighting by day as well as night—and notice if it is a fine day or if it's morning; and when they arrive at work they can play the standard charade, by saying how bad the weather is for this time of year. They can purse their lips as though they have just come in from an icy winter rather than from a heated car. They can contemplate the elements, as the elements are still called, from a bland distance. The crocuses, daffodils, and bluebells of spring can still lift hearts, even when seen only through a glass darkly by those who do not have gardens to potter about in. The red and yellow leaves fluttering down from another autumn can still surround an old memory with nostalgia. People can even notice that "lighting-up time" changes every day, and, if questioned, may recognize that their headlights are shining in obedience to the orbit of the earth. But winter and night can now be observed without having to conform to them. People get up and go to bed at more or less the same time the year round. Their lives are more uniform. With natural cycles held off, people can surrender themselves all the more fully to linearity, making it seem real by building a great scaffolding of measurement around the Tower of Time.

I have been describing one of the chief strands in social evolution which has proceeded by a series of steps, one building on another, from the Babylonians to the modern mass media. The framework of time belongs as much to the now common culture of the world as does technology in general. Unless a good part of it had been there beforehand, the industrial revolution would not have happened, and in due course added further to its sophistication. The new world society may not have any government of an ordinary sort; but in a human attempt to approach the sun, more successful than Icarus, we have built a single giant clock which tells the millennia and the centuries, the years and the hours, the minutes and the seconds for the world it hangs over and dominates. In its court there is no appeal.

Time has also been linearized by modern scholarship, modern technology, and modern counternatural organization. With all these devices, people devote themselves to squeezing more events onto the imagined line. Regularity is still there in the new habits, but for most people the day is less regular. Particularly if people are better-off, they see different acquaintances each day. They travel. Their leisure is more varied, with golf taking up some time, or collecting Napoleonic helmets or Roman denarii or models of the boats which have won the Fastnet race. A great number of meetings have to be arranged deliberately in order to match one timetable with another. This can extend even to relationships with members of their own families. High-pressure businessmen may manage only because they marry wives who are prepared to subordinate their own plans to their husband's.[40] But children are not necesssarily so amenable, and a lot of arrangement may still be necessary to make sure that husbands and wives are both at home when the guests arrive, or both get to the theater at the right time, or know each other's movements at all.

The more deliberation there is, the more deliberation is necessary. The less behavior is governed by routine, the more time has to be spent managing the time which can be less and less taken for granted in a society which has become obsessed by it. In this situation each new day or year resembles the previous day or year a little less than it did before. The successive changes with any lasting value are kneaded into what is already there and incorporated into the cyclical, which contributes the continuity to society. But the cycles are a little less straight repetitions than they were. They embody marginally more of the linear.

The increase in the scale of organization would not have happened with space unless it had also happened with time. The common framework consists of a concatenation of informal and formal agreements, founded on the increasing sophistication of measurement and secured, above all, by the greater willingness of people to undergo entrainment of their time habits than entrainment of any others. The deep propensity for cycles of which I have spoken makes it comparatively easy for the time controllers of the metronomic society to entrain others in their own interests. They can trade on a fundamental amenability. If people are told they have to start work at eight o'clock or secondary school at the age of eleven, or are to have the same length of working day in winter as in summer, they ordinarily obey rather than cavil. The common temporal framework of society rests upon a consent that hardly has to be asked for. We can get a sense of how arbitrary it is by consulting those who have not yet been plunged into the new kind of clockstream. The "nonchalant indifference to the passage of time" of the Kabyle peasant may not be so archaic after all. If the unification of mankind can be achieved only by making time so scarce that increasing numbers of people no longer have time to take wise decisions about its course, perhaps it will never be achieved.

All Days Are Not Equal

What is a man,
If his chief good and market of his time
Be but to sleep and feed? A beast, no more.
Sure he that made us with such large discourse,
Looking before and after, gave us not
That capability and godlike reason
To fust in us unus'd.

Hamlet, 4.4.33–39

Time is immanent in everything, pervasive, omnipresent, the model for every deity. You cannot catch it because you already have it: in writing a book on time there has never been a moment when my subject was not with me, even if I was not always with it. It has been as close to me as my skin, or, because no skin can wall out time, even closer than that. We do not tell the time, time tells us, because, as Kant said, "all representations, whether they have or have not external things for their objects, still in themselves, as determinations of the mind, belong to our internal state; and because this internal state is subject to the formal condition of the internal intuition, that is, to time—time is a condition *a priori* of all phenomena whatsoever—the immediate condition of all internal, and thereby the mediate condition of all external phenomena."[1] I have not been denying that; but while accepting that time is at the foundation of all our representations I have argued that we can and do regard time as made up of two elements, the cyclical and the linear. I have been contrasting them throughout the book—in the first half giving more play to the cyclical, in the second half to the linear—and now in this summing up I want to try and bring them together.

As for the linear, in a sense it always contains the cyclical within it, and this is apparent even when we are most caught up in the cyclical. We can persuade ourselves that it is only the grand procession of the heavens, and that alone, which lightens the evening sky in the spring; in London, the city of trees, turns the new blossoms into confetti which speckles every car into a wedding car; makes Nurse Bellamy and the other night nurses put on their summer uniforms; and sends the ice-cream vans trundling along suburban streets with their amplified bells, and makes children who a year ago could hardly toddle run out joyously to catch up with the sound. The wind cannot be seen but it can be inferred from its effects (which is why the Holy Ghost can be conceived of as a wind), and when the wind makes a different music in the trees, it is because it can pluck the new leaves of spring. It is all as though it has often happened before. But we also know that no summer or spring is ever quite like another. Our children are ourselves again and also their own selves. The oaks and the redwoods die and are replaced by different trees. Uniqueness, though never complete, is at least omnipresent.

We therefore must acknowledge not only that the present is elastic, containing the past and the future, but—an even more striking fact of our consciousness—that it is always passing. This at least is a constant: as comprehended in our human time the present continuously emerges from what was the future and is consigned to what has become the past, in the linear manner that decisively prevents any moment from being the same as another. If the I in us is aware of a relative permanence from one moment to another, it is also aware that the permanence is only relative to the impermanence. Every instant dies.

> Never, never again
> This moment, never
> These slow ripples
> Across smooth water,
> Never again these
> Clouds white and grey
> In sky sharp crystalline
> Blue as the tern's cry
> Shrill in light air
> Salt from the ocean
> Sweet from flowers.[2]

As human beings, we are designed for consciousness of the ever moving present, even if we also have memories which enable us to ponder a present now gone and imagination to wonder what may happen in a future which (except when we are bored) we can hope is not winging its way toward us.

The perishability of the present and its ubiquity make only too plausible the view taken by physicists. Current cosmology extends the Darwinian linear account of the origin of species (without necessarily being influenced by it) much farther back into another linear account of the origin of the universe. Theists believe that God started the clock ticking: Saint Augustine said that "you, my God, are the creator of all creation, and if we mean the whole of creation when we speak of heaven and earth, I unreservedly say that before he made heaven and earth, God made nothing."[3] The physicists say that our universe was set in motion, in what Herschel called the "laboratories of the universe," and the arrow of time given its one-way direction by one great explosion (or decompression).[4] The apparent motionlessness of the stars outside the solar system is an illusion. The universe is in a state of violent explosion, in which the great islands of stars known as galaxies are rushing apart at speeds approaching the speed of light.[5] The Big Bang set these galaxies into their recessions; formed the planets and set them in their orbits; and on our earth, and perhaps elsewhere, propelled life along its evolutionary path. Among the debris from that first great explosion are the same magic lattice of the New York skyline at night, Hymn 162, modern sociology, blue whales, Japanese clocks that go ping on the hour, and our human concepts of time. The physicists' account has not yet been made into a creation myth like the creation myths of the great religions, and perhaps it never will; but it is an equally dramatic story.

According to this view, space-time has unfolded and is continuing to do so in a linear manner; everything that happens within the expanding universe, and so within the kaleidoscopic parish we occupy, is transpiring inside that same unfolding. We did not set the process in motion, and its unwinding does not depend on us. As surely as the waves echoed in the cave before it was discovered and the tree fell noisily even when there was no one to hear it, time passes even when there is no one around to register its passing. Time is unwinding for everything else as well as for us, so that we can share the same now. We have been wonderfully designed to grapple ourselves onto the continuously moving present with the greatest of ease and the confidence of commanding all that the eye can see spread out below, able to delight in the sparkling world which our memory has allowed us to detach from time to see and wonder over.

Conceived thus, the mind has not been the prime mover. I recognize that the two-way relationship between the human and social sense of the passage of time and the unfolding of space-time in the Einsteinian universe is still a mystery. But although the prime mover is outside us, as a crucial condition of survival our minds have been enabled, by this mighty internal construction, to latch onto the more general linear progression which has in its grip all forms of matter and life and which takes us back, in every present moment, to the beginning of the universe.[6] We are continuously responding to linear time as well as to the more domestic cyclical time of the solar system. It is only within that overall linearity that the motions of the earth in our solar system, suspended halfway between life and death, have through their recurrences created a certain temporal stability. Animate beings have modeled themselves on this inanimate planetary system and, aided by the energy and the timing of the sun, created a new kind of order. Life succeeds in elongating the present by making it recur; all forms of life can then support one another in their characteristic but different slow motions.[7]

In their mutual relationship the linear encloses the cyclical more than the other way around. The homely solar cycles which are mirrored in the microcosmic rhythms of society and the even more microcosmic rhythms of the self are not the outermost boundary of our consciousness; they are inside the expanding universe, which is on its way to the last present, when time will stop. Meanwhile (and what a meanwhile!) life can seek to reverse the process by making the past flow back into the present. The future slows down into the present and the present slows down into the past. Life is bound to move with the arrow. No living thing can exist anywhere except in the now, the same now as for all other creatures who rejoice in the great unfolding. But into that moving now a great deal of the past is admitted. We can read the evolution of the universe, of the earth, and of life on earth because so much of the past is with us in the present. Because the speed of light is limited, light reveals to us in the now what "was" far away as well as what "is" nearby: we can look through our telescopes into the past and see the way the universe was when the light reaching us now was first emitted, thousands, millions, and billions of years ago, at least for events "occurring close enough so that a ray of light would have had time to reach us since the beginning of the universe."[8] The record of the past of the universe is like the geological record of the earth's past contained in rocks and fossils. It is accessible to us in the present and does not depend on cyclical behavior. But there is also a great deal of the past which would not be recoverable without that distinctive means of endowing the present.

The past can guide us members of the recovery team about what to do in the future. The best gift from the past is the memory that we have a future; we can remember, on the saddleback between past and future where we are forever poised with nothing but our experience of the past and our imaginings of the future to guide us, that we have some opportunity to decide what to do next. By making decisions we can attempt to cut into the continuous process by which the future becomes the past.[9]

Three Cyclical Recoveries

We would not be able to demonstrate that the linear is so fundamental, through the approaches adopted by either biologists or physicists, without the cyclical. They have extended the past out of the present by reconstructing the past and admitting their reconstruction into the present, doing at a more abstract level what each one of us does continuously by creating continuities in the little processions of our lives. If Napoleon was right to say you can recapture lost ground but not lost time, it would deprive the cyclical of its role as the conservator of the past. By compressing billions of years into an instant the cyclical creates a layered simultaneity of the linear past in the linear present. The past does not move into decent obscurity; it is constantly recycled. A simultaneous recapitulation is effected in the same present moment, in a hospital ward or anywhere else, by the messages from the past carried by the three cycles I have distinguished—by the genes, by habit, and by memory.

There is some similarity among the three of them. The genes are the principal instrument for summing up what has worked before, since life first appeared billions of years ago, around the time the earth's molten crust solidified.[10] People are not immortal, but their genes are more nearly so. It is genes which have internalized the sun. The cyclical changes brought about by the motions of the moon and earth are so fundamental and regular a part of our environment that living creatures have had to accommodate themselves to them and make use of them. The rhythms were built into the creatures who were our ancestors and are built into us so firmly that the beats pulse away inside us even when we are shielded from the light which connects our internal clocks, the sun, and the Big Bang. Today's sun entrains the tendency laid down many suns and many moons ago. The whole multiple recapitulation of the past is both tenacious and fragile. The past lives on only by being repeated constantly. If the cycles of becoming were interrupted only for a moment, the whole biosphere would stop its celestial breathing and a billion years of evolution would be lost. It could happen: as the sun reaches down into the oceans where life started, the submariners are waiting enclosed in their nuclear machines to end it, by erasing all connections with the past. The genes are not quite as safe as Dawkins thought, inside their lumbering robots, inside their own submarines, communicating with the outside world only indirectly:

> Four thousand million years on, what was to be the fate of the ancient replicators? They did not die out, for they are past masters of the survival arts. But do not look for them floating loose in the sea; they gave up that cavalier freedom long ago. Now they swarm in huge colonies, safe inside gigantic lumbering robots, sealed off from the outside world, communicating with it by tortuous indirect routes and manipulating it by remote control. They are in you and in me; they created us, body and mind; and their preservation is the ultimate rationale for our existence.[11]

The past and present masters underpin the circadian cycles, and I have argued that habits, the second main system of retrieval, also belong to a genetic process. The tendency to form them was presumably there at the beginning of the human species; habits have imparted a measure of stability to every social structure. Whether or not habits have a genetic underpinning, they gain their durability by passing from one carrier to another in a long chain. Samuel Butler said that a chicken is simply the egg's way of producing another egg; likewise, a society is habit's way of perpetuating itself. The particular bundle of habits possessed by any individual, however, is of recent history. An individual is not born with any particular set of responses to the tendency, as he or she is born with a set of genes, but acquires them from the environment. Habits therefore embody what has been learned since birth; they have no longer a life than an individual but incorporate what other individuals have learned in the past. Although not an adherent of social Darwinism I am enough one of social Lamarckism to recognize that acquired habits are transmitted by means of social inheritance. Lamarck believed in the inheritance of acquired characteristics, and so do I if the inheritance of specific social characteristics is not considered to be genetic. However idiosyncratically, we learn our own habits from the habits of others, which are (or were) the habits of yet others. They are embodied in custom, tradition, and culture; some of their content can be traced back if not as far as the beginning of the human race then almost to the beginning of human society. Although the greater part of culture is of recent invention some is much more ancient. We know something about the *Iliad* and the Bible and the *Bhagavad Gita* and the *Analects*; we do not know who first controlled fire to tame the Big Bang or first domesticated the wild dog or the wild horse.

The third system, memory, formerly incorporated less of the past than the other two retrievers, because it was limited to what people could remember of their own lives and what they had been told orally about lives before theirs. Parents and others still pass on to their children their individual recollections and the memoirs of their families by word of mouth. Fewer parents are now the Homers of their own hearth, and memory has been supplemented not only by the long-standing method of habit but, more and more, by written, pictorial, and electronic records, which contain a special kind of collective memory. This collective memory is constantly reinterpreted and reconstructed in the same way as individual memory. The beauty of the past is that it is so much more malleable than the present. John Ziman said that "a scientific library is not primarily a quarry, nor a factory, but a store. It is the 'memory' in which each item is continually being rewritten as new results are transferred."[12] The more general shared memory is the same. No longer limited to the memories of individuals and a few of their ancestors, it now takes us back fifteen billion years to the beginning of the universe. It is also subject to the same continuous reinterpretation. The problem is that the collective memory is growing at a faster and faster pace—also exploding, one might say—without any increase in the capacity of the minds that have to take it in. It is still an open question whether culture can enlarge its scope in order to relieve the burden on the collective memory in the same way that habit offers relief to the individual memory. If culture cannot carry more of the burden of precious messages from the past and from written and other records, and individual memories relatively less, we could all be defeated by information overload. Only culture can teach us what to forget.

The Strategic Balance

The three cyclical systems fit together, and indeed make up one overall system, because each of them is a means of securing the recurrence of a series of living processes. Every process in the body, down to the most microcosmic scale, is perpetuated by genetic reproduction. The behavior of the whole body, and the person within it, is also perpetuated by means of habituation and memory, and these too have a genetic base to them; and the same tendencies maintain some permanence in the still greater whole of society. It is a combination powerful enough to prevent the past being washed away by the running rush of the present. Viewed in this light, the social structure is not at all the gossamer affair that it is sometimes portrayed to be. It has strengths which are all the greater because they are unseen. It may seem odd to claim that faith in a religion or a code of conduct in science, in the English common law or the United States Constitution, rests on habits of mind with a genetic base to them. But however much it may go against the grain of modern thinking to admit that some of the triumphs of human reason are buttressed by semiautomatic forces, at least it has to be accepted that the triumphs are likely to be more lasting if they are. The loss of coherence and stability in modern society would be happening much faster if our society were not supported by the same three ancient cyclical systems.

The three systems are all conservative and successfully conservative because they do more than perpetuate what has been. They all allow for change. The triple cycles all allow for the linear. The genetic system is subject to mutations, not all of which are necessarily wiped out. No habits are irremovably fixed and at the margin old habits can be replaced by new habits all the more readily because the survival of the old habits allows conscious attention to be concentrated on what some of the old habits are no longer appropriate for. Likewise, our individual and collective memories maintain a record of the past which is not only being added to but also being constantly remade.

Taken together, the three living systems in maintaining a balance between persistence and change are more complementary than in conflict. There has to be persistence if the lessons of past experience are to be embodied in present practice and there has to be a capacity to adapt to ceaseless changes in the environment. Even if no changes were introduced from outside society the restless striving of its members would introduce them from within. Persistence and flexibility are the two great recipes for stability in society or any other living structure. But there is a third recipe which is just as important. The balance between the unchanging and the changing has to be conducive to the survival of the whole. This balance in society is not self-regulating; if it is not viable, no compensating forces bring about corrections except those that may be summoned by human reason.

Many societies of the past have been extinguished because their persistences were too marked and their ability to change not marked enough. The same thing is happening to some contemporary tribal societies: they are disappearing. But the general danger is now the other way around, all the more so because in our culture it is difficult to accept that there can be too much change. The main argument of this book has been that habits have decisive advantages for the individual in the relatively effortless way they preserve what is serviceable from the past; and that the like advantages of culture are even more decisive. The whole body of our social habits preserves (by partially removing from the domain of reason) what is serviceable from the past and creates many individual habits from the collective, just as individual habits also contribute to the collective in a mutual reciprocity. Our culture gives us what we have in common as members of a society and so makes it possible for us to be together, work together, live together, love together. What we have in common is precious; but if it is to persist we need to guard against acting only according to what we conceive to be the freedom of our individual and collective wills, and recognize that changes need to be gradual if they are to augment rather than disrupt. The changes that last are those that pile one increment upon another in a drawn-out process, as life sprung forth and then clung so tenaciously and so inventively to the surface of a planet which, if it was more hospitable than any other in our solar system, was still highly inhospitable; as the earlier forms of life elaborated themselves into the present ones, including us; and as human society has become ever more complex without breaking up. The lesson is that extraordinary transformations can be accomplished if the individual changes of which they are made are gradual. If the change is to endure it has to be cumulated bit by bit, and the new kneaded into the old, whether in subsystems like the law or science or in cultures and societies taken as wholes.

The gradualness is necessary because that is how change is brought about in habits and customs of mind. We cannot afford to get out of touch with them, even though it is so easy to do so when we are unaware of them. We have to be blind (or partially blind) to our own deep, unconscious structures of recurrence, both those derived from the fixations of our own childhoods and the repetitions of our later lives and those derived from the general culture. We can be aware only in a general way that "the psyche has depth downward but it also extends backwards and through time so that somehow history is latently contained and unconsciously expressed in each individual."[13] Because we are in good part unaware of our individual and social habits we are not going to regard them as assets when they are revealed. We have to be unaware if our consciousness is not to be overburdened. We have to be blind if we are to accept without too much questioning values which, although necessary for consensus in society, cannot be arrived at and upheld by the exercise of reason alone. We "feel" rather than "think" our values. It is in a sense necessary that we believe the right choices can be made by taking account of no more than what is in our conscious minds, in the light of the reason which is apparently our only faithful guide. We have to suffer from a kind of structural arrogance and ignorance. All societies (and persons) are liable to fall victims to this arrogance and, in deference to what seems the most compelling common sense, to put hastily on one side regularities which have taken thousands of years to establish.

Tradition has defenses of which the conscious mind is ignorant. Innovations are secretly sieved before they are given an audience; even if adopted they may be allowed only a short life before they are abandoned as incompatible with the underlying drift. But the defenses are not as impenetrable as they were, and the sieves are less fine. I gave pride of place to the accelerating advances in technology as the driving force behind social evolution. These advances have been supported by a cluster of new habits which are fully compatible with technology—I called them anti-habits. They uphold the practice of novelty and the enhancement of certain kinds of freedom. The alliance of technology with these habits has certainly been powerful, aided by the injection into the cultural combination of new time measurements. The alliance has completed the voyage begun by Columbus by throwing a web across the world, and bids fair to evolve a new interthinking species. I regard this outcome as a form of progress, but under challenge when novelty is so much prized and freedom of choice so much revered. Culture is being asked largely to desert its role as the conservator of society and devote itself to its other role as innovator. The switch has not happened yet. Society continues to hold together because the proportion between what is hidden away in the deep structures and what floats on their surface is not necessarily so different from what it was when almost the whole force of the culture was devoted to upholding what had stood the test of time. But that proportion is changing and not to the advantage of the more durable.

The current crisis in the family shows how precarious in one respect the situation is. The family is at once the most traditional of social institutions and the most essential, responsible for transmitting to children the clusters of consistent habits which give them security enough to be capable of love, capable of forming new habits as they mature, and capable of flexibility. The stability of the family matters so much because it is the model for behavior in all other institutions; its stability has not been exactly enhanced by having to cope with streams of new friends and neighbors in a more mobile society, the media which present ever different ways of behaving, the promise of novelty offered by new sexual partners. If impatience with the traditional becomes ever more de rigueur, more than the family will be destabilized, perhaps irredeemably. If it is now less true that "that which has been is that which shall be, and that which has been done is that which shall be done, and there is no new thing under the sun" (Ecclesiastes 1:9) the family may have been the chief sufferer.

The family, though far and away the most important, is only one of the general-purpose institutions like the community and the church faring badly in the face of specialized institutions which have taken away so many of their functions. Although society as a whole with its behavior and its culture remains the main overarching institution, the importance of specialization is growing all the time. Social evolution has consisted, like natural evolution, of ever more specialized features which have been integrated into ever more complex wholes. The threat to this integration has become steadily more severe. The diversity of the more complex structure has developed faster than the new kinds of holism that are needed. Almost everything in modern society is segregated—religion, education, art, death. The echoes of a Rome in decline, which have so often sounded throughout history, return again, as more people surrender restlessly to exotic doctrines and strange

sects because their traditional beliefs no longer give enough meaning to their lives.[14] There is not yet an outer, if permeable, envelope
for the emerging world society whose global character makes
humankind all the more vulnerable when there are hardly any
alternative societies left, and not exactly general acceptance of the
view that "in the search for the meaning of life in the modern
world, very little can be accomplished until we understand at least
the first axiom—that it is only out of the distant depths of his
being that man may find what the future holds in store."[15]

Replacing Nature

But if social evolution of the modern kind poses such dangers it
does not need to proceed in the same way in future. We could
gradually change its direction. We could reintroduce more of the
cyclical into the evolutionary mix and, particularly, give a more
prominent place once again to the cycles of nature. Their replacement on a large scale by human-made ones has been as recent as
it has been far-reaching. An equally radical earlier change in the
methods of production, the Neolithic revolution, altered the
rhythms of life without breaking their tie with nature. The new
agriculturalists had to conform to the cycles of the day and the
seasons, and to the life cycles of the animals they domesticated,
as faithfully as the hunters and gatherers who preceded them. On
the other hand the industrial revolution has set up rhythms which
are opposed to those of nature. The millions of people who have
been caught up in it have left the land to take up a new life inside
buildings with an internal climate, internal seasons, and internal
lighting. The everyday environment has become an indoor environment with square corners to it. Mechanisms akin to genetic
reproduction have been persuaded to replicate anything, even the
motions necessary to drill a spotface in a cycle of 2.059 minutes.
   The new attitude is that man is in control, not nature. The
instruction in Genesis 1:27–28 has been taken literally: "God said
unto them, be fruitful, and multiply, and replenish the earth and
subdue it: and have dominion over the fish of the sea, and over
the fowl of the air, and over every living thing that moveth upon
the earth." So enjoined, one limb of nature has managed with

brilliant success to control and subdue the rest, and "all extensively colonized land has become to some degree a human artefact."[16] The outcome has been pollution of the skies, the seas, the rivers, and the land by toxic chemicals and noise, the extinction of one species after another, the exhaustion of natural resources and the growing scarcity of time. We have forgotten that if we are the lords of creation we rule only by consent.

We would not have been able to think of controlling time had we not been able to measure it with increasing precision, dividing it into many different units exactly equal in size. The units are constant as long as we have counting machines within sight to tell them off for us. This measurement of time has underpinned science since Pythagoras and underpinned technology since the invention of the clock escapement, first in China and then in Europe. The first of the new tools essential for the industrial revolution would not have emerged from the Boulton and Watt workshop in Birmingham without the new proficiencies in measurement. The division of labor has depended throughout on different subsidiary parts fitting closely together to make a whole. Mass production has required hundreds and thousands of components to be so identical and therefore interchangeable that each Rover automobile, or Sony television, or IBM computer produced from a factory is measurably the same as all the others. The mechanical and electronic cycles which fine measurements make possible are in their finest manifestations almost as exact in their recurrences as those described by the spin of the earth. They depend on impressing metronomic time into human organization as well. People are made-to-measure like other factors of production. In and out of car factories their working days are measured, determining when they should start and stop, go for lunch, have a tea break, or take their holidays; how much they shall be paid, down to the last integer; and, always, what they should do with their time. The precise divisions which the mechanical clock has added to the astronomical clock have been used to make interchangeable those who do the most brutally repetitive of society's work, while the precise parceling up of the life cycle has created the career anxieties which have been used to control the salariat.

This hyper-measurement has also made time into a neutered

"it," in place of the anthropomorphic figures of the past—Father Time, Chronos, the Reaper—who controlled us far more than we controlled them. Time is commonly represented by a spatial metaphor so that we can put our ordinary senses (and particularly our eyes) to work on its mystery and so try to make up for the fact that natural evolution has given us no organ of time perception as it has given us sight and touch for objects in space, hearing for sounds, smell for scents, and taste for food. Time is as invisible, untouchable, inaudible, unsmellable, untasteable as the idea of God, and as indissolubly part of everything. Only the metaphors which are time's language allow us to think otherwise.

Yet metaphors are so apt for time precisely because they cannot be tested by reference to our ordinary senses. We cannot consult them to ask if time is really like a reaper or a whirlwind, a sergeant or a siren, a bird in flight able to stay aloft against the wind only by beating its wings or a grain of sand in an hourglass. Metaphors can therefore become the thing itself; although the product of human minds they can so reify time that it seems a fact of nature, an absolute, existing in its own right. "Reification implies that man is capable of forgetting his own authorship of the human world."[17] A common property of the metaphors is that they separate this metaphorical time from us who invented the metaphors. A stock example is the river of time—which superficially resembles linear time insofar as it is always flowing, and only in one direction (as long as one forgets that the water in the river returns as rain to its source). According to the modern view we are on the banks, watching the river flow toward us and then away into the distance behind us or walking alongside toward the water ahead as we leave other water behind us. Either way, with watch in hand, we can divide the river upstream of us into durations such as tomorrow, next month, or next century, which we or our descendants shall in due course occupy and make use of; or the river downstream into yesterday, last month, or the last century, which no one will occupy again. Because we stand beside the river, we can observe time and measure it and imagine we can control it like some irrigation engineer. The metaphor is more comforting than if it threw us into the river, to be borne along toward the ocean without being able to get out. We would then lose our watches in the water.

As it is, we can pretend, with the aid of riverproof clocks which seem to be measuring time from outside it, that the modern world has discovered the solution to Saint Augustine's problem. We can delude ourselves that we know what time is because we know what time it is. If clocks can measure time, as in the modern view they axiomatically can, we can summon all our stores of confidence to pretend that we understand what we can measure. The unfortunate corollary of our faith in measurement, and its increasingly sophisticated application to both time and money, is that we do not value what cannot be measured, evidenced by the way we do not value education unless it can be made subservient to measurements culled from tests and examinations. We do not value the mystery of time or of anything else unless we think we can bite into it with the pincers of measurement.

Another even more unfortunate consequence is that we do not have time for each other, unless it is time that is parceled into the right segments, and in human care how many needs can be parceled out in that way? The metronomic processes which have been so effective in industry and in the delivery of some services have proved ineffective when it comes to care. "Despite the inroad the new technologies have made already or are going to make into the home, community and family life, smiles cannot as yet be exchanged via computer, nor can the warmth of a hand, nor the non-verbal support in interpersonal relations be easily portioned, stored or dispersed at fixed intervals."[18] The most signal characteristic of the family, and of friendship which goes beyond exchanges of convenience, is that the time which it makes available is in principle unlimited, provided according to need and not according to the expectation of a like return. Familiarity breeds content if it is not metered. Children need almost unlimited time, as do many old people and sick people, but increasingly they do not get it. They get their measured meed instead of what they need. Money is no substitute for care; and care loses whenever it has to be calculated or weighed out—posing the most severe problem for the welfare state and for individual relationships.

In the task of learning more about the new ascendancy of the linear—trying to unpack an element from the general culture in order to examine it—sociology could play a distinctive part. As part and parcel of a developing temporal awareness, a chrono-sociology (with or without that name) is needed alongside chron-obiology to study the rhythms and the habits of society; there is also a need for a temporal psychology to go hand in hand with the new sociology and the new biology. Implicit in my use of the three main analogies is my belief that none of the existing disciplines should stand alone. Time does not lend itself to specialization. Time is not a subject like other academic subjects. Nor is social evolution. Increasingly biologists will need to become sociologists and psychologists, as sociologists will need to become biologists and psychologists, and psychologists the other two, all with as much interchange as possible with historians and philosophers; and for any of the students to ignore physics, the queen of the sciences, would be to ignore the field in which the largest advance has been made so far in the conceptualizing of time. All the disciplines, and new hybrids between them, will be required to join in the coming unification of the human and the natural sciences in the next century; all in order to increase the understanding of themselves by individuals and societies without destroying the balance between what is understood and what does not need to be understood.

Back to Circadian Rhythms

Meanwhile, I can only argue my case for adhering more fully to natural rhythms by reference to what we already know. I described earlier how the manifestations of biological clocks in sleep and digestive cycles are almost as formative in industrial societies as in others, but I did not consider what happens when the clocks are out of order or their requirements ignored by modern social arrangements. The importance of the biological clocks is like that of other bodily functions. We are not aware we have kidneys, livers, or biological clocks until they go wrong; and then we become only too acutely conscious of how much we rely on them to be uncomplainingly regular. "Our body," said Robert Burton in 1628, "is like a clock; if one wheel be amiss, all the rest are disordered, the whole fabric suffers; with such admirable art and harmony is a man composed."[19]

Mental disorders. People with certain mental disorders have biological clocks which have clearly gone out of adjustment. For instance, sufferers from what was identified over a century ago as *la folie circulaire* seem to have permanently destabilized clocks. Sometimes called cyclopaths, these persons experience episodes of mania and depression that recur in a double pattern, with daily rhythms in the severity of their symptoms superimposed on longer-term cycles. Durkheim was one of the first to notice the seasonal cycles in suicide.[20] They are liable to follow seasonal cycles in depression, with the highest incidence in the spring, and with another smaller peak in the autumn.[21]

In these disorders biological rhythms get out of kilter with each other, perhaps because the biological clocks cannot coordinate other internal rhythms or the rhythmic release of hormonal or chemical messengers. The switch from depression to mania—in which feeling flat, pointless, and shapeless gives way to a sense of ease, power, and well-being, and then to overwhelming confusion—often starts with a sleepless night. The patient moves temporarily from a 24-hour sleep-wake cycle to a 48-hour one. Treatments based on the manipulation both of sleep (by retiming the hours when it occurs) and of light have shown promise.[22]

Temporal distortions may underlie many other mental disorders as well. Jacob Arlow is a psychoanalyst who has studied these matters. He has shown that in a state of depression people are liable to feel defeated by the past, and in a state of anxiety to feel that they are bound to be defeated by the future.[23] The psychiatrist F. T. Melges claims that some schizophrenics typically suffer from a breakdown in their power of segmented thinking.[24] They cannot keep track of sequence and so cannot formulate plans of action for themselves. According to him, there is hardly a neurotic or psychotic disorder in which the relationship to time is not out of balance. If he is right, we are not exactly in control of that relationship, the undisputed masters of our fate, so much as dependent upon the rhythmic balance being right.

Toxicity of drugs. Things can go seriously wrong if the circadian rhythms of people ill from mental or other disorders who are being treated with drugs are ignored. This is because people's reactions to drugs vary according to the time of day. Many drugs are, for instance, more effective and have fewer side effects when administered at certain times of day. If the time of day is right, the chances of recovery can be considerably improved. To give one example, a recent study has shown how much timing matters in chemotherapy for cancer.[25] Drugs such as adriamycin and cisplatin can kill not only cancer cells but also normal cells. The susceptibility of normal cells, however, varies rhythmically during the circadian cycle much more than that of malignant tissue. The study showed that a group of women were worse off when adriamycin was administered in the evening and cisplatin in the morning than those in a group for which it was the other way around. In the first group twice as many patients had to have reductions in dosage, four times as many treatments had to be delayed, and drug dosages had to be modified three times as often.

Over cyclical reactions like these we as individuals have little or no control. We can only hope that, if we are ill, we can persuade those who care for us to be aware of our circadian sensitivities. On other matters we may have some choice, both about whether to depart from social cycles which are fully in accord with the biological ones and about the steps to moderate the harmful consequences if we do.

Jet lag. Although we can self-confidently adjust our wristwatches on a transatlantic flight, anticipating New York time as soon as the wheels leave the ground at Heathrow, we cannot do so with our biological clocks. They follow their own rules, reminding us again that we are after all like other terrestrial animals.[26] Our internal clocks obstinately continue to tell Greenwich Mean Time after our bodies and our wristwatches have arrived in New York.

In this respect people are not so different from bees. Some cycles lag in their adjustment more than others until, after some days, the system is restabilized by the time signals of the new environment. Adjustment to a longer day after westward flights is generally easier than to a shorter day after eastward flights: the circadian system can move more readily toward its natural free-running period of 25 hours than against that tendency. Whichever the direction of travel, it appears that the social cues of the destination help reset the internal clock. The best policy seems to be to immerse yourself right away in the new social round of your destination, its meal times, working times, noises, smells, hours of sleep. Even if you do, the adjustment does not happen immediately, and for some people it is very difficult.

Night work. People cannot order themselves to work at night and expect that their bodies will unresistingly comply. They may find it a little easier if they are "owls" or "evening types" whose cycles are delayed compared to the rest of us. They can better manage to sleep, even if they do not get the chance to lie down until others are getting up. They also adjust more rapidly to a westward flight. "Larks" or "morning types" are better suited to day work and to eastward flights. We cannot choose which to be, or control the change from one to the other at different stages of our lives; but whichever we are, continuous night work is a risky business.

People who choose to stay up at night, like the fishermen and women on the Venice pier, are one thing; people who have to work at night under the same pressures as in the day are another. They are worse off than transmeridian travelers. If they are on a weekly rotation, they have to put up with a reorientation every week equivalent to a long flight. Very rarely can they subordinate everything else in their lives to night work as though ordinary days did not exist. Their rhythms are continually re-entrained to days by the ordinary pressures of social life, and because this is normal for their bodies the effect is immediate, whereas the adjustment to night work can be long and drawn out and must be gone through each time the schedule resumes.

Yet the number of workers on shifts has been increasing; in both Britain and the United States something like one quarter of the workforce is on "unusual" hours.[27] As technology advances and capital equipment becomes more expensive, there is more and more pressure to keep costs down by operating around the clock. The number of people on nights in service occupations has also shown no sign of falling off. One reason is the shrinking of the world of commerce so that there is more communication across time zones. New York currency brokers open as early as 5:30 a.m. to handle business with London and have a second shift which carries on until midnight to deal with Tokyo.

Evidence has been steadily accumulating about the harmful consequences. Social and family life are liable to impoverishment, sleep to interference and health to strain. There is an increased incidence of digestive disorders and gastrointestinal complaints, which are unlikely to be due just to disturbances in eating habits. Judgment is generally impaired in the middle of the night. Doctors, air traffic controllers, pilots, nuclear submarine crews, and operators in nuclear plants sometimes have to make vital decisions when they are least capable. The nuclear accident at Three Mile Island began at 4 a.m., the one at Chernobyl at 1:30 a.m., and the explosion at Union Carbide in Bhopal, India, at 12:40 a.m. There is a case for cutting down on night work or, where it is essential and workers get paid more for it and have a choice about whether to do it, for giving them better information so that those who do choose it know what they are letting themselves in for. They might be less willing if they knew more about how their bodies worked.[28]

If night work has to be done, the damage can at least be minimized by a sympathetic rotation. A good example comes from the work of C. A. Czeisler and his colleagues in Boston.[29] They were able to show improvement in factory workers' feelings of good health, reduction of labor turnover, and gains in productivity when a rotation incorporating "phase delays" replaced one incorporating "phase advances." The rotation had been to follow a week on a midnight-to-8 a.m. shift with one from 4 p.m. to midnight and then one from 8 a.m. to 4 p.m. The researchers recommended introducing a phase-delayed rotation, with a longer period for adjustment on each shift before changeover. The new weekly rotation started with midnight to 8 a.m. and moved "backward" to 8 a.m. to 4 p.m. and finally to 4 p.m. to midnight. The changes from week to week were thus in the same direction as the internal drift; whenever people free-run they move by means of phase delay toward the slower cycle of 25 hours. Phase delays can therefore be entrained much more readily than phase advances. As anticipated, the workers adjusted more rapidly from week to week on the new rotation, just as people recover more quickly after westward flights. The same technique has been used with success to treat some kinds of insomnia.

Monotonous work. The same question can be raised about monotonous work as in other instances of social arrangements not being in accord with biological cycles. For most people work is not monotonous when it follows the internal circadian rhythm, that is, with regularities from one day to another, but it can be when the regularities are within the day. Monotony is likely to set in (although of course it does not always have to) when cycles are reduced to 35 minutes, as in the school, or even more to 2 minutes, as in the factory. Such human-made cycles, which exist in every organization, do not correspond with circadian ones; if they correspond with an ultradian cycle it is only by coincidence. Such short cycles seem to be contrary to nature. Dancing and other rhythmic movements can be in time with the heartbeat, but there are very few organizational cycles in which the directors and other employees dance their way through the day.

Furthermore, these very short work routines are often at the bottom of a vast hierarchy of cycles. The low-paid housekeepers of society in the home are joined by people at the bottom of every ladder who have to march to a quicker beat. Writing in the 1940s, two London doctors, I. H. Pearse and L. H. Crocker, founders of the Peckham Health Centre, said of the weekly wage earner:

> He expects to have to fit himself into other people's convenience in all that he does. His hours of work are inexorably determined for him, he is not free to arrive an hour earlier and leave an hour earlier as is his master; if he wants a job, or hospital treatment, he must attend at some hour determined by those who have either at their command. If his wife wants to see the child's teacher or the school doctor she must attend at some stated time not necessarily convenient to her. Nowhere in fact—except by the private dentist or hairdresser—is the time and convenience of the weekly wage-earner and his family considered as anything to be respected.[30]

The change over forty years, at least in Britain, has hardly altered that picture. The movement toward greater equality—and flextime has helped a little—has barely touched blue-collar workers; by and large it is only white-collar workers who find it easier to take time off and to use the phone for personal calls. In the places where I have done research the ordinary employees have a great deal of routine within the day, and within the week, month, and year for that matter. Managers and executives like Mr. Murgatroyd are not on 2-minute cycles, although at one remove they are still geared to those below them whose regularity they initiate, monitor, and perpetuate.[31] The foreman's bosses are even further removed from the shop floor. The regularity in managers' schedules is over longer periods related to the annual budget and corporate plans and the timetables associated with them. Managers' salaries are often calculated not on hourly, weekly, or monthly terms, but on an annual basis, with expectations of increase related to a lifetime career structure. Their daily lives may be marked by the Urgent driving out the Important, but most managers strive against this pull of the immediate and try to control the organizational prime time, that is, the longer term, the more distant horizons. If they do that successfully, they will also control the shorter term of those whose jobs do not bestow such time privileges.

Research on the consequences of the different styles of life, particularly for health, has shown that in certain circumstances those at the top suffer more from coronary heart disease and other complaints. This is true at any rate of those exhibiting "Type A behavior"—ambitious, competitive, important people, who speak rapidly and are accustomed to a rapid pace of activities.[32] Such people, engaged in a "chronic incessant struggle to achieve more and more in less and less time, and if required to do so, against the opposing efforts of other things or other persons," were said to be especially at risk.[33] In several European studies this finding has not been borne out, and the emphasis has instead been on the risks of people with relatively low status. M. G. Marmot, in a long-term study of British civil servants, demonstrated that although there was more Type A behavior in the higher grades, mortality was much greater in the lower: "Compared with the highest grade (administrators) men in the lowest grade had 3 times the mortality rate from coronary heart disease, from a range of other causes, and from all causes combined."[34]

Why should this be so? One factor may be the greater monotony which goes with the repetitiveness of the shorter cycles. The extent to which people can control the pace of their own work also matters, the worst-off being those with short cycles controlled by machines and by the managers whom machines can call to their aid. The extent to which monotony oppresses people, and their productivity at work, appear to vary considerably in different countries depending on the individual control the workers exercise, how long their countries have been industrialized, the political climate, and other variables. In general, it is as disagreeable for most people to be controlled by machines as it is agreeable to control them. People who are allowed to pace themselves can change the pace when they are feeling the strain, or perhaps stop for a rest. Learning new things on the job also reduces monotony.[35] Many routine workers, in manual and in nonmanual jobs, seem to be understimulated rather than overstimulated, with too little pressure on their working time rather than too much, which can be as oppressive as overstimulation.[36] Prisons impose inactivity as a punishment. Understimulus can give rise to resentment, to ill health, and perhaps, if protracted, even to death. Boredom is one of the greatest problems of a technological society in which more and more of the work is done by people who are using less and less of their potential for responding innovatively to lives which have too little challenge in them.

At all events, there is not only agreement inside every organization about the recurrences to be followed and the privileges attached to certain recurrences, but, as is to be expected if the distribution of the congenial is so unequal, a good deal of conflict as well. As I tried to show in my description of the factory, many workers devote themselves to trying to beat the system. It is done everywhere, not just by varying the pace of work in a manner disapproved of by the time-study men, but in dozens of other ways. In a school all of the children some of the time, and some of the children all of the time, are joined in battle with the teachers to secure the joy of even a token release from the regular. In all organizations transgressions and departures may be acceptable if they are confined within close limits, but the extent to which this is allowed is small, and will in retrospect seem distressingly so if the modern drummer who keeps people marching in the new lockstep becomes progressively less insistent. Time truants are still penalized, excluded, dismissed.

Reversing Evolution

The five examples above are obviously not conclusive, but they raise warnings about the damage which, in our arrogance about time, we may be doing to human beings whose internal constitutions were established long before machines were thought of. Social evolution may have followed natural evolution and yet be in conflict with it. At any rate my argument is that it is perilous to override our biological rhythms, which are tied in to astronomical rhythms, as we have done and are increasingly doing. This must seem like flying in the face of social (if not natural) evolution, after all that I have said about it and particularly after I have given a role to technology which seems to require that people (or at any rate the majority of people) should submit fully to the discipline of the mechanical clock and other machines. What I propose is actually a reversal of social evolution, not wholesale but at the margin. There are precedents for such reversals in nature: "There is no reason why general trends in evolution should not be reversed. If there is a trend towards large antlers for a while in evolution, there can be a subsequent trend towards smaller antlers again."[37] With the help of the anti-habit we could nudge the trend in a different direction; we surely do not need to wait until much larger numbers of people have taken to time-bending drugs to try and escape from the new locksteps.

The first point on the agenda for such a change is a new attitude toward death, which would emphasize its cyclicity more and its linearity less. We know from the experience of others—all our knowledge of death comes from society—that death belongs to the natural rhythm of the life cycle. But death has become more difficult to accept as more features of human society appear to have been brought under control without making the ultimate threat any less decisive. Mortality has been postponed but not eliminated in every industrial society by better food, hygiene, housing, and medical treatments, and the physical pain of its approach perhaps reduced. Although life is quintessentially cyclical in that genes, habits, and memories are passed on from each older to each younger generation, each life also comes to a stop. A life is therefore also quintessentially linear.

If a life were not linear the reproductive cycles would not be needed to act as bridges between the past and the present. But death ensures that the multiple connections over time between one generation and another are not too close. If they were, the adaptability of human beings, and all other forms of life, would be greatly weakened. Imagine (as so many people have) that death were eliminated. Would we be here at all to speculate about it? If the individual members of any species were everlasting there would be no evolution. Evolution depends on more offspring being born than will survive or reproduce themselves. Those with harmful mutations and recombinations could not be eliminated without death as the selective agent. Moreover, human beings would have to pay a special penalty; if we were all Methuselahs, habits and memories would not be subject to the continuous revolution of the generations. Children cannot copy their parents' habits exactly. Just as children acquire recombinations of the parental genes, so do they acquire and construct new combinations of parental habits and those of others around them. Without the benefit of death, memories would not be restated and revised whenever a new person takes the place of an old. By making room for variations in the copies, death encourages the variability and adaptability which sustain life. Montaigne justified the ways of death by saying, "Make room for others, as others have done for you." They are also justified because the others will not be the same as you.

Perhaps death will not be necessary if instability becomes more of a problem than stability. But for the time being, and I would expect for a long time to come, mortality will remain 100 percent. This is the only tyranny which is not just unqualified but justifiable. The only attitude in the end is to accept it rather than rage against it, as can be done with other tyrannies less inescapable, for life would eventually stop in all species if individual lives did not. Without individual death, collective death would in the end be inevitable. It is not only soldiers who die for others but all of us. Death enforces altruism. Individual mortality ensures a collective immortality that has become ever more comprehensive with the enlargement of the collective memory.

We should accept the sentence of death not just for the sake of others. The character of Edgar in *King Lear* comments that "Men must endure / Their going hence, even as their coming hither: / Ripeness is all." It is also not just for endurance's sake. As individuals, we gain some consolations to offset our prospective loss; although stoicism about oneself is easy compared to stoicism about others whom we find out how much we need when we can no longer tell them. But life would stop being the gift it is if it lasted forever—witness the trouble people have always had in making their heavens delectable without the edge provided by death. Life is precious because it partakes of the great linear unfolding, not just in its end but in the way each moment within it ends and thus reminds us of how a life will end. While it is still in the future each moment, as it emerges in any day in any life, is as much an unknown as death itself. A life mimics the universe in every moment as well as over its whole course. If every moment were not so transitory, what would happen to the savor? How would we avoid another kind of death, from the extremity of boredom? We all play games with time in the same manner as the three great cycles I have been describing. We go on improvising the games which make up what we are by freezing time, like lovers with eternal longings in our eyes, like poets who hark to the lark so attentively it seems to sing forever, like artists who paint such brilliant poppies they are in full flower as long as the canvas lasts, like men who plant trees to make forests. While dicing with our everyday epic we also know that the acceptance of death, in a person or a moment, is what makes us aware of the preciousness of life.

Time is our most precious possession and one we have in common. When we recognize its dread but not unrelenting scarcity we know that our joys are joys partly because they are shaded. Joy has to be poignant. If it were not, we would have less reason to agree with Saint Augustine that "every particle of sand in the glass of time is precious to me." Technology may have added to the poignancy rather than moderated it: more particles in the glass of time, and more abundances in the parts of the world more favored by new technologies, make it seem there is more to lose. Dr. Johnson, when he saw the lovely furniture, the books, and the beautiful women around David Garrick, said "Davey, Davey, this must make death very terrible."[38] We are not all Garricks, but death is bound to be daunting partly because we are not allowed any practice at experiencing it, except vicariously from others who cannot tell.

The prospect would be more daunting without it, however, even for individuals for whom the interests of the species as a whole do not bulk all that large, at any rate when their own interest in life is under threat. If the absence of death would be even more daunting than its presence, perhaps in due course more people will be able to recognize and understand its place within the rhythms of nature. Likewise, we need to be conscious of the scarcity of time rather than denying it, although the consciousness has to be kept in check. If it is too insistent life can lose its savor, not because nothing much moves but because everything is moving too fast. The happy mean seems to be to remain aware of the passage of time but to behave tranquilly as though one is not aware of it and has, at least on some occasions, all the time in the world.[39] There are different degrees of what Freud called the "oceanic feeling"—the notion of limitless extension and oneness with the universe—and all are meaningful. Time moving gently can generate a sense of unity with all life, and beyond life, with the universe.

The second, much lighter, point on the agenda—reacquaintance with the seasonal cycles—is not quite so difficult. The life cycle cannot be set aside in any society, whereas the seasons to some extent can. It has been done before in imagination—the Garden of Eden as portrayed in the Bible has no seasons—and it has now been done for real. Perhaps not all of Adam's or Eve's descendants regard the buffering of climate as having brought us nearer to Paradise; I don't. To pay more attention to the seasons, especially at latitudes away from the equator, could be as therapeutic as to ritualize more openly the revolutions of the earth. We could also become more aware of the great unfolding of the universe which is speeding us along with the most distant galaxies which are invisible even to our modern superenhanced vision, as it has carried along all the people, from the creators of Stonehenge to the moon landers, who have been driven to explore their relationship to the scudding dome above them. We may not be able to feel on our pulses the unfolding of the universe but we could try to open our eyes and ears and nostrils to the unfolding of each year even if we are not ourselves like daffodils that take "the winds of March with beauty."

Escape from the Grid

If we wish to copy nature without copying it exactly, to learn from it without being slavish, the third point on the agenda has to be the day, on which the new metronomic society has so fully imposed its gridiron. In this respect society seems to be becoming more, not less, amenable to Dickens' "deadly-statistical clock . . . which measured every second with a beat like a rap upon a coffin-lid."[40] A theme that has run through this book is that clocklike regularity is not the way of nature, and if we want modern habits to be more in tune with nature, we will have to follow the archetypal model more closely. Its lesson is variety, including spontaneous one-off variety. I saw a slogan painted on a Mammy wagon in Ghana that summed it up: "All Days Are Not Equal." In nature, no one minute is the same as another; the play of light changes, the shadows lengthen. No day is the same as another; its length changes and it provides so many other surprises—a snowfall, a storm, a rainbow, the rain that does or does not fall. There is a constant change of pace, with moments of intensity followed by relaxation. While at one minute after six with bent heads the clerks shuffle homeward down into the tube station, above their heads, if only they would look up, the starlings are excitably wheeling in the sky and the swifts darting about in the fading light to remind them of what was and could be. The swifts are not killing time or being killed by it. They are taking advantage of a dusk which is not yet a night. Coleridge said of another latitude, "the stars rush out: / At one stride came the dark." But farther from the equator such changes are gradual, from dawn in a hospital to dusk above a London tube station, or transitions from one season to another. The abrupt changes modern people are required to make come not from nature but from the metronomic society.

There is no precedent in nature for the car factory, or any factory, office, or school, any more than there is for us creatures who are so obsessed with one kind of time; but there is precedent for the alterations in pace that still occur, even inside the buildings where people live and work. Edward Thompson believes that alternate bouts of intense labor and idleness constitute a natural rhythm of work.[41] This rhythm is followed by many self-employed people such as artists and writers, and even small farmers and craftsmen who if they are dragooned at all are not dragooned by others. The workers in Mr. Murgatroyd's car factory strove for a similar rhythm, and more and more others could in the future. Mental fatigue as well as physical fatigue can be brought on by any task which calls for continuous short-cycle repetition, without the relief and the stimulus that a change of pace can bring.

The same purpose would be served by loosening up formal organizations to withstand the regularities of the machine and the corporation. The main target should be the short cycles within the day. But it would be no use going to the other extreme: praying for nothing but variety would be like praying for nothing but rain. The proper balance is a matter of searching for different relationships between the rhythmic and the linear, and recognizing that no one can know what fits someone else. People do not necessarily know what suits themselves either; in the course of overriding nature, habit causes a good deal of suffering as well as bringing so many benefits. But individuals are far more likely to know what suits them than society does. The more people pick their own rhythms, the better the chance they will pick the right ones. Society would not survive everyone's making their own choices, nor would many people survive having to; but at least there do not need to be so few choices for so few about how to make use of their most precious asset.

Behind the other issues is the issue of measurement. Should we cling to the natural, or to the uneasy blend with the natural that modern human beings have introduced with so much ingenuity and effort? If we proceed to add yet more precision to the metronome and go on requiring people to conform to it even when it runs against their nature, always demanding that they turn about when the calendar tells them they have reached a certain age or the clock a certain time, and calling on them to divide the future into neat-looking parcels which they can imagine they are controlling, we will aggravate the hurry sickness.[42] But if we would become less grim in our determination to measure up our tomorrows for our tomorrow's selves, as though fearing we would feel lost without a straitjacket; if we would let the sense of time refer to something besides being on the dot, and always knowing where the dot is; if we would leave more room for spontaneity; if more of us could let the future move into the present without grasping at it, with more presence of mind, the time famine could begin to abate.

So much counting within the stop-start life cycle, within the year, within the week, within the day and within the hour, even within the minute, has created the time-bound society. Any minute can be made into a beginning or an end for thousands of people. We are continually having to start this or stop that at the appointed moment. In a society always looking at its watch we have to listen for the sergeant who commands us to not be "late" for the start of any event, or to leave "early" before it has come to its term, or to change step just because we have passed a particular but arbitrarily chosen birthday. Life is strung between a whole series of precisely timed beginnings and ends which have a lot to do with man and little with nature. Outside our building, the transitions from darkness to light, and from winter to spring, are slow; if people can become part of them, they can become aware of forces more fundamental and more calming than the mechanical overlay they have so diligently clamped down on themselves. Fast food, fast cars, fast work, fast talk, fast news, fast bucks, fast sex will all be self-defeating until patience is again made into a virtue and time made into the principal dimension of human ecology.

Am I just crying for the moon—and the sun? Did not my argument turn on the peculiar thrust that the growth of technology has given to human evolution? And does not technology require that people be placed under the strict discipline of the clock and have punctuality drummed into them at school and demanded of them thereafter? Can people have the standard of life which they have become habituated to desire along with the luxury of living as though their contract is with nature? I can say no more than that technology does not have to remain the prime mover. Machines, instead of being masters, can also be liberators, freeing people not just from work which is harsh by reason of its physical arduousness but from work which is harsh by reason of its temporal imperatives. Machines could enslave themselves to their own master machine, the clock, so that we should no longer be required to treat ourselves as though we were machines. *They* could hum in *their* metronomic society, leaving us free in our own more human rhythmic society to set out on the next stage of evolution, which will be more different from anything that has gone before than anything that has gone before. As Dennis Potter has said, the past is always running alongside us. The shadow runner is larger than life, but we can turn the direction in which we are running and take the shadow with us.

New possibilities could open up, not of returning to a world we seem to have lost but of recovering that past in a new form, offering new human concertos made up of rhythms within rhythms, of daylongs and yearlongs that both mimic and vary our lifelongs, and lives that recall from the fiery passage of time the resonating memories that make it meaningful in ever new ways. In the next stage we could remove some of the elaborate cloaks in which time has been hidden, look more closely at some of the taken-for-granted methods we have used to try to tame time, and in a new era to which time is central make altogether less of the scarcities of space and be more bold with the scarcities of time. We could move toward a freedom from the clock and re-establish an entente with the sun. A new approach to time could be the key to a new enlightenment as we try to understand, in order to reveal without resolving the mystery which has been at the heart of every religion; in awe, but not overawed by whatever it is which is moving the future into the present and the present into the past; in the supreme hope that the question about the potential of our nature, as individuals and in society, and its relationship to the vaster nature of our universe, will never be finally answered.

Epilogue

It is dawn at an English West Country fishing port on a summer morning. Norman Widdicombe, the skipper of the *Mary Jane*, has just been woken by an alarm clock which is reset each night to a different time. His wife no longer hears it. Her life is hard enough without having to get up when he does: "She has to stay at home listening to the wind all day."

Still in his pajamas, he opens the bedroom window so that he can get a better view of the harbor, where his trawler is waiting for him and his two crew members. A draft of air rushes in, unseasonably cold for the time of year. He taps the barometer on the wall by the window. It has dropped during the night. It will be no use going out to sea if there is to be a gale; he could lose his nets, and more. Norman's father was a skipper too and he remembers again what his father told him when he was first taken fishing as a boy: "It is better to be at home wishing you were at sea than at sea wishing you were at home." Mistakes about the weather are best made on land.

So give it another quarter of an hour. He makes himself a cup of tea and puts on a sweater. When he next looks out the red and brown colors of the hulls of the fishing fleet can be picked out from the greyness, and the whiteness of a few yachts. The fish-market building has a firmer outline. He strains to focus with his binoculars on the waves beyond the outer breakwater. They have white tops. He can just see the seagulls flying around his mast.

But it is a Wednesday. If he goes out this morning he will be able to stay out two days and a night on the fishing grounds to save fuel and still be back on Thursday night in time for the market on Friday. There is no market on Saturday, so he cannot wait a day. It is also August; there should be red mullet and squid to catch in mid-Channel, although it means going eighty miles. This is the season for mullet as they migrate from their spawning to their feeding grounds and swim, far out, past his window. All the fish which give him his livelihood pass up and down the English Channel in their seasons.

It should be a good night. The mullets and squid like swimming against a strong tide. There should be plenty of them, and he knows just where to find them. The prices for mullet and squid have been good this week, especially on the French market. His own takings have been poor. He needs the money from a good catch.

He telephones David and John to say he's going down to the boat and will get the ice. He will wait on board for the six o'clock BBC news and weather forecast, and sniff the wind before finally deciding whether to sail this day.

Notes

1. The Cyclical and the Linear

1. J. Adams, S. Folkard, and M. Young, "Coping Strategies Used by Nurses on Night Duty," *Ergonomics*, 29 (February 1986), 188–196.

2. M. C. Moore-Ede, F. R. Sulzman, and C. A. Fuller, *The Clocks That Time Us* (Cambridge, Mass.: Harvard University Press, 1982), p. 81. According to the Koran dawn has arrived when you can see the difference between a black thread and a white thread.

3. John Locke, *An Essay concerning Human Understanding* (London: Dent, Everyman's Library, 1961), I, 166.

4. Bernard Leach, *The Potter's Challenge*, ed. David Outerbridge (London: Souvenir Press, 1976), p. 19.

5. Mircea Eliade, *The Myth of the Eternal Return, or Cosmos and History*, trans. W. R. Trask, Bollinger Series, 46 (Princeton: Princeton University Press, 1954), pp. 113–114.

6. Ibid., p. 18.

7. Christianity has been perhaps even more influential as a force for linearity, Christ's birth and crucifixion being marvelously unique events. Not that in this respect the Judeo-Christian tradition is on its own: "If one demands something still more numinous, the life of the Sage, the Teacher of Ten Thousand Generations, Confucius (Khung Chhiu, 552–479 B.C.), supreme ethical molder of Chinese civilization, the uncrowned emperor, whose influence is vitally alive today in the tenements of Singapore as well as the communes of Shangtung, forming the inescapable background of the Chinese mind whether traditional, technical or Marxist; this life was at least as historical as that of Jesus." J. Needham, "Time and Knowledge in China and the West," in J. T. Fraser, ed., *The Voices of Time* (London: Allen Lane, 1968), p. 134.

8. J. E. S. Thompson, *The Rise and Fall of Maya Civilization*, 2nd ed. (Norman: University of Oklahoma Press, 1967), p. 167. See also G. J. Whitrow, *What is Time?* (London: Thames and Hudson, 1972), chap. 1.

9. Some modern societies are not all that dissimilar. Clifford Geertz, in *The Interpretation of Cultures* (New York: Basic Books, 1973), has described how the "permutational" calendar, built around the interaction of independent cycles of day-names, has worked in modern Bali. On p. 128 he quotes P. Radin, *Primitive Man as a Philosopher* (New York, 1957), p. 227, on an Oglala (Sioux) view: "The Oglala believe the circle to be sacred because the great spirit caused everything in nature to be round except stone. Stone is the implement of destruction. The sun and the sky, the earth and the moon are round like a shield, though the sky is deep like a bowl. Everything that breathes is round like the stem of a plant. Since the great spirit has caused everything to be round mankind should look upon the circle as sacred, for it is the symbol of all things in nature except stone." Marc Bloch has suggested that the Balinese calendar does not rule out a linear view as well as a cyclical or "nondurational" notion of time. M. Bloch, "The Past and the Present in the Present," *Man*, 12 (August 1977), 284.

10. For Nietzsche the readiness to accept eternal recurrence is the test of strength, and not just in ancient societies. In *The Gay Science* he says, "What if some day or night a demon were to steal after you into your loneliest loneliness and say to you: 'This life as you now live it and have lived it, you will have to live once more and innumerable times more; and there will be nothing new in it, but every pain and every joy and every thought and sigh and everything unutterably small or great in your life will have to return to you, all in the same succession and sequence—even this spider and this moonlight between the trees, and even this moment and I myself. The eternal hourglass of existence is turned upside down again and again, and you with it, speck of dust!' Would you not throw yourself down and gnash your teeth and curse the demon who spoke thus? . . . Or how well disposed would you have to become to yourself and to life to *crave nothing more fervently* than this ultimate eternal confirmation and seal?" Friedrich Nietzsche, *The Gay Science* (New York: Vintage, 1974), book 4, p. 273.

11. An earlier discussion of terminology is in M. Young and J. Ziman, "Cycles in Social Behaviour," *Nature*, 229: 5280 (1971), 91–95.

12. E. R. Leach, "Cronus and Chronos," in *Rethinking Anthropology* (London: Athlone Press, 1961), p. 125.

13. *Van Gogh: Vincent by Himself*, ed. B. Bernard (London: Orbis, 1985), p. 204.

14. As De Quincey said, "Space swelled and was amplified to an extent of unutterable infinity. This, however, did not disturb me so much as the vast expansion of time; I sometimes seemed to have lived for 70 or 100 years in one night; nay, someimes had feelings representative of a millenium passed in that time, or, however, of a duration far beyond the limits of any human experience." Thomas De Quincey, *Confessions of an Opium Eater and Other Writings* (Oxford: Oxford University Press, 1985), p. 68. We have to try to accept despite its

difficulties for modern minds that the unconscious (and the recycling that goes on through it) is not a negative concept. "One of the great paradoxes which we must learn to comprehend is the fact that it is precisely when we feel most conscious and at our highest point of rationality that non-conscious forces penetrate the area of awareness and either becloud or dominate consciousness." I. Progoff, *Jung's Psychology and Its Social Meaning* (London: Routledge and Kegan Paul, 1953), p. vi. This applies as much to the expressions of the social unconscious of culture as to the individual unconscious.

15. J. Needham says that in ancient China elixirs were "time-controlling substances." See "Time and Knowledge in China and the West," in *The Voices of Time,* ed. Fraser. The slowing down or the heightening of consciousness may represent the same kind of action as when a dolphin speeds up its rate of emitting clicks whose echoes enable it to navigate from "cruising rate" to 400 clicks per second as it is closing in on its prey. R. Dawkins, *The Blind Watchmaker* (London: Longman, 1986), p. 96.

16. E. Zerubavel, *Patterns of Time in Hospital Life* (Chicago: University of Chicago Press, 1979).

17. Roland Fisher, "Biological Time," in *The Voices of Time,* ed. Fraser, p. 360. According to this account time perception changed with the decreasing rate of metabolism.

18. This was once described in the following way by a bakery worker who was very bored because he had one of the easiest jobs: "It seems like there's some bugger standing on the hands of the clock and stopping them going round." J. Ditton, "Absent at Work; or How to Manage Monotony," *New Society,* 22 (21 December 1972), 679–681.

19. William James, *The Principles of Psychology* (New York: Dover, 1950; first published in 1890 by Henry Holt and Company), I, 624.

20. Aristotle, *Physics,* book 4, chap. 10, 217b33. Quoted in R. Sorabji, *Time, Creation, and the Continuum: Theories in Antiquity and the Early Middle Ages* (London: Duckworth, 1983), p. 8.

21. Saint Augustine's paradoxes of time are discussed by Susan Wilson et al. in *Time* (Milton Keynes, England: Open University Press, 1973), which leans on J. J. C. Smart, *Problems of Space and Time: A Reader* (London: Collier Macmillan, 1964).

22. Saint Augustine, *Confessions,* trans. R. S. Pine-Coffin (London: Penguin Books, 1961), p. 269. Whitehead had another way of expressing the same idea when he said that "what we perceive as present is the vivid fringe of memory tinged with anticipation." A. N. Whitehead, *The Concept of Nature* (Cambridge: Cambridge University Press, 1920), p. 73.

23. James, *Principles of Psychology,* I, 609–610.

24. Liam Hudson, in an excellent book, says that this is quite different from the petrifaction which occurs when art becomes someone's property. "Such petrifaction by ownership or patronage is, of course, different from that inherent to the act of painting itself; but is perfectly compatible with it." Liam Hudson, *Human Beings: The Psychology of Human Experience* (London: Jonathan Cape, 1975), p. 195.

25. *The Timaeus of Plato,* translated and with a running commentary by Francis MacDonald Cornford (London: Routledge and Kegan Paul, 1937), pp. 97–98. "When the father who had begotten it saw it set in motion and alive, a shrine brought into being for the everlasting gods, he rejoiced and being well pleased he took thought to make it yet more like its pattern. So as that pattern is the Living Being that is forever existent, he sought to make this universe also like it, so far as might be, in that respect. Now the nature of that Living Being was eternal, and this character it was impossible to confer in full completeness on the generated thing. But he took thought to make, as it were, a moving likeness of eternity; and, at the same time that he ordered the Heaven, he made, of eternity that abides in unity, an everlasting likeness moving according to number—that to which we have given the name Time."

26. Locke, *Essay concerning Human Understanding,* p. 166.

27. Sebastian de Grazia, *Of Time, Work, and Leisure* (New York: The Twentieth Century Fund, 1962), p. 318.

28. Ernest Gellner, *Thought and Change* (London: Weidenfeld and Nicolson, 1964), p. 48.

29. John Dewey, *Human Nature and Conduct* (New York: Holt, 1922). On this point Richard Wollheim has made splendid use of the entry in Kierkegaard's *Journal*—"It is perfectly true, as philosophers say, that life must be understood backwards. But they forget the other proposition, that it must be lived forwards." I do not think there is now any danger of forgetting the other proposition. R. Wollheim, *The Thread of Life* (Cambridge, Mass.: Harvard University Press, 1984).

30. A leading psychologist has described what happens even in very young children: "A one-year-old, for example, is shown eighteen different pictures containing a pair of dogs of different shapes, sizes and colors, and looks equally long at each of the dogs. On the nineteenth trial, the infant is shown a new dog, but one paired with a bird, and looks much longer at the bird than at the dog." Jerome Kagan, *The Nature of the Child* (New York: Basic Books, 1984), p. 237.

31. R. L. Gregory, *Eye and Brain: The Psychology of Seeing,* 3d ed. (London: Weidenfeld and Nicolson, 1977), chap. 11.

32. Saint Augustine, *Confessions,* p. 264.

33. E. O. Wilson, *Sociobiology* (Cambridge, Mass.: Harvard University Press, 1975), p. 575.

34. Anthony Giddens, *Central Problems in Social Theory* (London: Macmillan, 1979), pp. 7–8.

35. Giddens has said on this point, "The characteristic view of the synchronic / diachronic distinction is that to study a social system synchronically is to take a sort of 'timeless snapshot' of it. Abstracting from time, we can identify functional relations, how the various contributing elements of a social system are connected with one another. When we study systems diachronically, on the other hand, we study how they change over time. But, the result of this is an elementary, though very consequential, error: *time becomes identified with social change.* One should notice that the synchronic / diachronic division presumes the Kantian dualism of space and time, the first being available for synchronic

analysis in abstraction from the second. However, it is more important in this context to stress the point that time (time-space) is obviously as necessary a component of social stability as it is of change . . . A stable order is one in which there is close similarity between how things are and how they used to be. This indicates how misleading it is to suppose that one can take a 'timeless snapshot' of a system as one can, say, take a real snapshot of the architecture of a building." A. Giddens, *A Contemporary Critique of Historical Materialism* (London: Macmillan, 1981), p. 17. My difference from Giddens in this passage is that I regard one sort of change (the cyclical) as being essential to social stability, and so do not contrast stability and change. I do identify time with social change, but with two markedly different kinds of social change. Giddens has provided an excellent treatment of time in "Time and Social Organization," in his *Social Theory and Modern Sociology* (Cambridge: Polity Press, 1987).

36. Isaiah Berlin, *Vico and Herder* (London: Hogarth Press, 1976), p. 64.

37. Pitirim Sorokin was ahead of his time (as they say) when he remarked in 1928 that "changes in social life for the last few decades; a failure of the eschatological conception of history and that of the attempts to discover *the* 'historical trends'; a better knowledge of many social phenomena; discoveries of many brilliant civilizations of the past; these, and many other factors, are responsible for the fact that social thought seems to begin again to pay a somewhat greater attention to the repetitions, rhythms and cycles in social and historical processes." P. Sorokin, *Contemporary Sociological Theory* (New York: Harper, 1928), p. 729.

2. Extraterrestrial Timers

1. J. D. Palmer, *An Introduction to Biological Rhythms* (London: Academic Press, 1976).

2. Ibid., p. 131.

3. Daniel J. Boorstin, *The Discoverers* (London: J. M. Dent, 1984), p. 42.

4. According to Milton, this is not an advantage: God put an end to Eden's perpetual spring when he commanded his angels to "turn askance / The poles of Earth twice ten degrees and more / From the sun's axle." *Paradise Lost*, ed. E. M. W. Tillyard (London: George Harrap, 1960), book 10, lines 668–670.

5. "I recall some very early experiments of the Nobel laureate, Otto Meyerhof, who was a biochemist, and who—just to play a long shot—used to put big bottles full of inorganic and organic chemicals on the shelf in his laboratory, sterilize them and let them sit there while he went about his business for the next 20 or 30 years. He was hoping, of course, that some day the miracle would occur, that suddenly some life would appear in one of his bottles." H. P. Klein, "Introductory Comments," in *Life in the Universe*, ed. J. Billingham (Cambridge, Mass.: MIT Press, 1981), p. 19. Anyone repeating Meyerhof might shorten the odds an iota by subjecting the bottles to light and dark cycles, or at least putting them outdoors rather than keeping them in the lab. The results actually achieved

by passing electric sparks and ultraviolet light through such flasks have been decidedly impressive, and they might be more so if the flasks were cycled.

6. See M. C. Moore-Ede, F. R. Sulzman, and C. A. Fuller, *The Clocks That Time Us* (Cambridge, Mass.: Harvard University Press, 1982), p. 3.

7. Circadian rhythms need to be distinguished from what have come to be called biorhythms (also known as bio-curves and bio-cycles). The biorhythm theory, developed by Wilhelm Fliess in the 1890s in Germany, holds that there are three cycles: a 23-day (physical or male) cycle that influences physical strength and endurance, a 28-day (emotional or female) cycle that influences feelings, sensitivity, and emotional reactions, and a 33-day intellectual cycle comprising changes in intelligence, alertness, or awareness. It pretends to be able to predict human behavior in terms of "bad" or "good" days at any time of a person's life. According to a recent enquiry there is no evidence that such cycles exist. K. E. Klein and H. M. Wegmann, *Significance of Circadian Rhythms in Aerospace Operations*, AGARDograph no. 247 (London: N.A.T.O. Advisory Group for Aerospace Research and Development, 1980).

8. J. L. Cloudsley-Thompson, *Biological Clocks: Their Function in Nature* (London: Weidenfeld and Nicolson, 1980), p. 15.

9. Palmer, *Introduction to Biological Rhythms*, p. 144.

10. Ibid., p. 149.

11. There is an excellent general account of these and other rhythms in D. S. Minors and J. M. Waterhouse, *Circadian Rhythms and the Human* (Bristol: John Wright, 1981).

12. J. Adams, S. Folkard, and M. Young, "Coping strategies used by nurses on night duty," *Ergonomics*, 29 (February 1986), 188–196. As another consequence of renal rhythms, kidney transplants into rats have been much more successful when done in the evening than in the day.

13. N. Kleitman, "Studies on the Physiology of Sleep," *American Journal of Physiology*, 104 (1933), 449–456. See also W. P. Colquhoun, ed., *Aspects of Human Efficiency: Diurnal Rhythm and Loss of Sleep* (London: English Universities Press, 1972).

14. R. J. Broughton says that adult human sleep has an inherent tendency "towards a bimodal expression about every 12h per day which, in normally entrained subjects, usually occurs as a major nocturnal sleep period and a briefer period of afternoon sleepiness or overt sleep." "Three Central Issues Concerning Ultradian Rhythms," in *Ultradian Rhythms in Physiology and Behavior*, eds. H. Schulz and P. Lavie (Berlin: Springer-Verlag, 1985), pp. 217–233.

15. S. Folkard and T. H. Monk, "Circadian Performance Rhythms," in *Hours of Work: Temporal Factors in Work Scheduling*, eds. S. Folkard and T. H. Monk (Chichester, England: Wiley, 1985). Folkard has found that some kinds of mental performances have a cycle of about 21 hours. The 21- and 24-hour cycles are therefore in phase every eight days. He suggests the week may have a physiological if not an astronomical tie.

16. H. Hoagland, "The Physiological Control of Judgements of Duration: Evidence for a Chemical Clock," *Journal of General Psychology*, 9 (1933), 267–287. Quoted in Palmer, *Introduction to Biological Rhythms*, p. 137.

17. Cloudsley-Thompson, *Biological Clocks*, p. 47.

18. Ibid., p. 34

19. Ibid., p. 36.

20. Ibid., p. 71.

21. Although it would be strange indeed if the correspondence between the menstrual and the lunar periodicities turned out to be accidental, the connection is not yet understood. It does seem that women can synchronize each other. A study in the University of Stirling in Scotland showed, over four months, a trend toward synchrony between women students and their closest friends, and suggested that the mechanism could have been the cycle in vaginal odors (without people being aware of it). C. A. Graham and W. C. McGrew, "Menstrual Synchrony in Female Undergraduates Living on a Coeducational Campus," *Psychoneuroendocrinology*, 5 (1980), 245–252. There is no obvious way in which moonlight or gravity could be implicated, with different phases of the moon acting as signals for different women. A possible explanation is that there could once have been a selective advantage in menstruation occurring as the moon began to wax. Women would then be in their most fertile periods while the full moon gave the best light for seeing their way out of the cave or camp. As reliance on natural light declined, the same length of period may have been retained but the onset of menstruation detached from the same phase in the lunar cycle.

22. Cloudsley-Thompson, *Biological Clocks*, p. 97. See also John Brady, *Biological Clocks*, Institute of Biology's Studies in Biology, no. 104 (London: Arnold, 1979), p. 14.

23. Cloudsley-Thompson, *Biological Clocks*, p. 31.

24. See J. E. Lovelock, *Gaia: A New Look at Life on Earth* (Oxford: Oxford University Press, 1979). Inanimate objects also play a part in atmospheric chemistry and its cycles. "All the elements upon which living organisms depend move in biogeochemical cycles, which carry them between the organisms themselves and the non-living parts of their environment, and the great majority of them must pass through the air at some stage in the cycle." M. Allaby and J. Lovelock, *The Greening of Mars* (London: Andre Deutsch, 1984), p. 62.

25. Brady, *Biological Clocks*, pp. 38–39.

26. Ibid., p. 40.

27. Moore-Ede et al., *The Clocks That Time Us*, p. 7.

28. Palmer, *Introduction to Biological Rhythms*, p. 76.

29. Ibid., p. 62.

30. Ibid., pp. 126–127, 128. See also J. Aschoff, "Circadian Rhythms in Man," *Science*, 148 (April–June 1965), 1427–1432.

31. Moore-Ede et al., *The Clocks That Time Us*, p. 87.

32. Gina Kolata, "Genes and Biological Clocks," *Science*, 230 (December 1985), 1151–1152.

33. Ibid., p. 1151.

34. Taken from A. S. Iberall, "New Thoughts on Bio Control," in *Towards a Theoretical Biology*, vol. 2, ed. C. H. Waddington (Edinburgh: Edinburgh University Press, 1969), p. 167.

35. Walter B. Cannon, *The Wisdom of the Body* (London: Kegan Paul, 1947), p. 24.

3. Timing of Social Behavior

1. M. Douglas and B. Isherwood, *The World of Goods* (London: Allen Lane, 1979), p. 119.

2. Frederick W. Taylor, *Scientific Management* (New York: Harper, 1911).

3. "Conflicts over time are only one route. Conflict arises over how work shall be organised, what work-pace shall be established, what conditions producers must labour under, what right workers shall enjoy, and how the various employees of the enterprise shall relate to each other." E. Edwards, *Contested Terrain: The Transformation of the Workplace in the Twentieth Century* (New York: Basic Books, 1979), p. 13.

4. Daniel Bell, *Work and Its Discontents* (Boston: Beacon Press, 1956), p. 16.

5. The sitting, as unnatural as the timing, may be responsible for much of the endemic back pain in adults. As one physiotherapist said, "At the age of five, a child in an industrialised or developed society begins to differ from its counterpart in a primitive environment . . . The modern urban child is obliged to adapt its basic programme to a system that demands that it should be imprisoned for hours at a time behind a school desk, which is more often than not ill-fitting and poorly designed . . . Many of these children leave school eight to ten years later with a ruined posture." C. Hayne, "Alive, alert, and 'primative,'" *Self-Health*, 12 (12 September 1986), 8–9.

6. I. Cullen and E. Phelps, "Diary Techniques and the Problems of Urban Life," Report to the British Social Science Research Council, no. HR.2336 (1975).

7. B.B.C. Audience Research Department, *The People's Activities* (British Broadcasting Corporation, 1965); and *The People's Activities and the Use of Time,* Annual Review of BBC Audience Research Findings, no. 5 (British Broadcasting Corporation, 1977–1978).

8. I. Cullen, "Notes on the Measurement of Routine," unpublished manuscript (1984).

9. Tristram Shandy's father took regularity even further: "I was begot in the night, betwixt the first *Sunday* and the first *Monday* in the month of *March,* in the year of our Lord one thousand seven hundred and eighteen. I am positive I was . . . My father . . . was, I believe, one of the most regular men in every thing that ever lived . . . he had made it a rule for many years of his life,— on the first *Sunday night* of every month throughout the whole year,—as certain as ever the *Sunday night* came,—to wind up a large house-clock which we had standing upon the back-stairs head, with his own hands:—And being somewhere between fifty and sixty years of age, at the time I have been speaking of,—he had likewise gradually brought some other little family concernments to the same period, in order, as he would often say to my uncle *Toby,* to get them all out of the way at one time, and be no more plagued and pester'd with them the

rest of the month." Laurence Sterne, *The Life and Opinions of Tristram Shandy, Gentleman* (Oxford: Clarendon Press, 1983), pp. 8–9.

10. See K. K. Sillitoe, *Planning for Leisure*, Government Social Survey (Her Majesty's Stationery Office, 1969), and H. B. Rodgers, *The Pilot National Recreation Survey*, report no. 1 (British Travel Association and University of Keele, 1967).

11. Michael Young and Peter Willmott, *The Symmetrical Family* (London: Routledge and Kegan Paul, 1973), p. 217.

12. H. L. Wilensky, "The Uneven Distribution of Leisure: The Impact of Economic Growth on 'Free Time,'" *Social Problems*, 9 (Summer 1961), 37. Quoted in Young and Willmott, *The Symmetrical Family*, p. 146.

13. J. Gershuny, "Growth, Social Innovation, and Time Use," in *The Economics of Human Betterment*, ed. K. E. Boulding (London: Macmillan, 1984), p. 55.

14. Ibid., p. 49. See also J. Gershuny et al., "Time Budgets: Preliminary Analyses of a National Survey," *Quarterly Journal of Social Affairs*, 2, no. 1 (1986), 13–39.

15. J. P. Robinson, *The Use of Time: Some Trends and Patterns in American Research* (Organisation for Economic Co-operation Development, 1978), p. 23.

16. Ibid., p. 25.

17. There are, of course, many different rhythms within any city. Jonathan Raban has said of two different quarters of London: "Time in Earl's Court is quite different from time in Islington. The north London square took its rhythms from the five-day week and the eight-to-six working day. Its shops opened at eight-thirty and closed at seven; and by day, women and children had the place to themselves. In Earl's Court, somewhere is open all the time; supermarkets do not close until midnight, and after that there are cafes, clubs and hot-dog stands . . . For the rest of the day and night, there is a continual tidal sweep of comers and goers, and they make their own suns and seasons, as if this was a completely synthesised underground city." *Soft City* (London: Hamish Hamilton, 1974), pp. 191–192.

18. T. Carlstein, *Time Resources, Society, and Ecology: On the Capacity for Human Interaction in Space and Time*, vol. 1, *Pre-Industrial Societies* (London: Allen and Unwin, 1982).

19. Leonardo da Vinci, *The Codex Hammer, 1508–1510* (London: Royal Academy of Arts, 1981), p. 46.

20. H. S. Bennett, *Life on the English Manor: A Study of Peasant Conditions, 1150–1400* (Cambridge: Cambridge University Press, 1937), p. 96.

21. There are still the seasonal anxieties and the accompanying folklore: "Will August storms flatten the wheat again? Will cutting drag on into September with the combine stuck axle deep in mud? How will we pay the grain-drying bill, let alone the wages? But it didn't rain on St. Swithun's Day, and the rooks nested high in the trees this year." P. Redgrove, "Holy Days: May Day at Padstow," in *About Time*, ed. C. Rawlence (London: Cape, 1985), p. 61.

22. Émile Durkheim, *Elementary Forms of the Religious Life*, 2nd ed. (London: Allen and Unwin, 1976), p. 378.

23. Daniel J. Boorstin, *The Discoverers* (London: J. M. Dent, 1984).

24. Ibid., p. 8.

25. W. Lloyd Warner, *The Family of God* (New Haven: Yale University Press, 1961), p. 361.

26. V. W. Turner, *The Drums of Affliction* (Oxford: Clarendon Press, 1968), pp. 269–270.

27. The modern calendar cannot compare with that of the Elizabethan Age, as it was rhapsodized over by R. H. Tawney: "The imaginative vivacity which, when fired by a crisis, produced the poetry of action, flowed, in the tranquil routine of normal life, through different channels, but was sustained and invigorated, not stifled, by it. In villages, a round of recurrent activities—May-games; Whitsun, Easter and Christmas festivities; Church-ales; yearly wakes; occasional 'gathering for Robin Hood,' such as, on one occasion, had deprived an indignant Latimer of his congregation; . . . these and similar diversions, if not universal, appear to have been widespread." "Social History and Literature," in *The Radical Tradition* (London: Allen and Unwin, 1964), pp. 203–204. A very different picture is given of Coventry in 1519 by Charles Phythian-Adams, *Desolation of a City: Coventry and the Urban Crisis of the Late Middle Ages* (Cambridge: Cambridge University Press, 1979), pp. 74–76: "when the only artificial illumination available to the mass of people was gained from firelight or tallow-dips, the length of the working day depended essentially on the hours of day-light, which naturally varied from winter to summer . . . From such beginnings, the medieval working day extended for twelve to fourteen hours. If the 1493 statute was observed the day was broken by set times for meals . . . half an hour was allowed for breakfast, an hour for 'dinner,' and half an hour for 'noon-meat.' Only between mid May and mid August was a siesta permitted, for which purpose the dinner-break was extended to $1\frac{1}{2}$ hours." In that city mastiffs were let loose to roam the streets by night, to serve a double purpose: to frighten the citizens into staying indoors in the dark hours when mischief could so easily be done, and to clean the streets by eating refuse. It is as if everyone were ordered off the streets of New York at night and thousand upon thousand of attack dogs let loose as the clocks struck curfew in order to terrorize people into observing it.

28. The linear element in sport is almost as notable as the cyclical. The excitement is often in the novelty, in the breaking of some record—the fastest 100 meters, the shortest or the longest tennis match, the greatest number of home runs, the fastest Derby, or being the first man to take 100 catches in Test cricket.

29. For this information I am grateful to Dr. John Fletcher, lecturer at the University of Aston and a specialist in medieval educational institutions.

30. The peaking of vacations has often been regretted. One writer about the European Community noted: "Hotels with their shutters presenting a hideous and neglected appearance; restaurants and cafes closed; services of all kinds suspended until the following spring; travel facilities much reduced or even suspended. Such under-utilization of course means that overhead costs have to be recouped on a very short period of the year, instead of being spread out over

12 months, with the inevitable consequences for the tourist of much higher prices for peak season usage. It is nothing short of a major scandal that tourists are exploited in this way . . . It has been established that some 40% of all persons taking holidays are free of these constraints, and therefore free to take their holidays when they like: but in practice even these people opt to join the crowds and to go on holiday during peak periods!" Ronald Martin, "The Staggering of Holidays in the European Community," in *Social Europe* (Commission of the European Communities Directorate-General for Employment, Social Affairs and Education, May 1984), pp. 76–77.

31. Eviatar Zerubavel, *The Seven Day Circle: The History and Meaning of the Week* (New York: Free Press, 1985), p. 12.

32. Ibid., p. 19.

33. See the section on the week in J. Goody, "Time-Social Organisation," in *Encyclopaedia of the Social Sciences,* ed. D. Sills (London: Macmillan, 1979), XVI, 33–34.

34. *Hansard* (House of Lords, London, 2 December 1985), col. 1070.

35. Zerubavel, *The Seven Day Circle,* p. 139.

36. R. R. Rindfuss and J. L. Ladinsky, *Medical Care* (1976), 14, 685. Quoted in A. Macfarlane, "Variations in Number of Births and Perinatal Mortality by Day of Week in England and Wales," *British Medical Journal* (1978), 1670–1673. Most of the facts I discuss come from this and other papers by Macfarlane.

37. It is no longer possible for so many children to enjoy the benefit they once got from a Sunday start, if the nursery rhyme is to be believed:

> Born of a Monday, fair in face,
> Born of a Tuesday, full of God's grace,
> Born of a Wednesday, merry and glad,
> Born of a Thursday, sour and sad,
> Born of a Friday, godly given,
> Born of a Saturday, work for your living,
> Born of a Sunday, never shall we want
> So there ends the week, and there's an end on't.

Iona and Peter Opie, *Oxford Dictionary of Nursery Rhymes* (Oxford: Oxford University Press, 1951), pp. 309–310.

38. A. Macfarlane, "Deaths: The Weekly Cycle," *Population Trends,* 7 (Spring 1977).

39. Anthony Giddens, *Central Problems in Social Theory* (London: Macmillan, 1979), p. 71.

40. Plato, *The Laws of Plato,* trans. A. E. Taylor (London: Dent, 1934), book 7, 809.

41. A. J. Casper and W. P. Fifer, "Of Human Bonding: Newborns Prefer Their Mother's Voices," *Science,* 208 (June 1980), 1174–1176. See also Gina Kolata, "Studying in the Womb," *Science,* 225 (July 1984), 302–303.

42. P. L. Harris, "Infant Cognition," in *Handbook of Child Psychology*, vol. 2, ed. M. Haith and J. Campos (New York: Wiley, 1983), p. 689. See also R. N. Aslin, D. B. Pisoni, and P. W. Jusczyk, "Auditory Development and Speech Perception in Infancy," in ibid.

43. Émile Durkheim, *The Division of Labor in Society* (New York: Free Press, 1964), p. 242; see also pp. 1—31.

4. Habit: The Flywheel of Society

1. John Dewey, *Human Nature and Conduct* (New York: Henry Holt and Company, 1922), p. 40.

2. This passage, along with several of the other quotations on which I draw, is reproduced in the masterly review by C. Camic, "The Matter of Habit," *American Journal of Sociology*, 91 (March 1986), 1051.

3. Émile Durkheim, "The Evolution and the Role of Secondary Education in France," in *Education and Sociology*, trans. S. D. Fox (Glencoe, Illinois: The Free Press, 1956), p. 152. Also quoted in Camic, "The Matter of Habit," p. 1052.

4. Max Weber, *The Theory of Social and Economic Organisation*, ed. T. Parsons (New York: The Free Press; London: Collier Macmillan, 1964), p. 116.

5. In psychology I have, in general, followed the position adopted by Liam Hudson in his *Human Beings: An Introduction to the Psychology of Human Experience* (London: Jonathan Cape, 1975).

6. See, for example, E. L. Thorndike, *Animal Intelligence Experimental Studies* (London: Macmillan, 1911; enlarged ed., New York, London: Hafner, 1965), pp. 68–69.

7. Anthony Giddens, *Central Problems in Social Theory* (London: Macmillan, 1979), p. 218.

8. The interest of Talcott Parsons, for instance, is in social action—action which is value-directed and based on an assessment of ends and means in the light of the norms which guide people. Social action is distinguished from behavior, which is in the domain of habit. Parsons in one passage leans on Weber: "It is clear that Weber directly associates the concept of action with an accessible subjective aspect, with the postulate of VERSTEHEN. In so far as human 'behaviour' (the broader category, VERHALTEN) is not accessible to such understanding through the subjective view of the actor it is not action and does not concern the formulation of Weber's systematic sociological theory." Talcott Parsons, *The Structure of Social Action* (Glencoe, Ill.: The Free Press, 1949), pp 640–641.

9. C. Camic, "The Matter of Habit," p. 1077.

10. Pierre Bourdieu, *Outline of a Theory of Practice*, trans. R. Nice (London: Cambridge University Press, 1977), pp. 82–83.

11. William James, *The Principles of Psychology* (New York: Dover, 1950; first published in 1890 by Henry Holt and Company), I, 121, 112.

12. Thomas Henry Huxley, *Lessons in Elementary Physiology* (London: Macmillan, 1866), lesson 12. Quoted in James, *Principles of Psychology*, I, 120.

13. Arthur Marwick, *Britain in the Century of Total War* (London: The Bodley Head, 1968), p. 161.

14. Henry David Thoreau, *Walden*, ed. J. Lyndon Shanley (Princeton, N.J.: Princeton University Press, 1971), p. 7.

15. H. Maudsley, *The Physiology of Mind* (London: Macmillan, 1876), p. 154. Quoted in James, *Principles of Psychology*, I, pp. 113–114.

16. James, *Principles of Psychology*, I, p. 122.

17. Jerome S. Bruner, *On Knowing: Essays for the Left Hand* (Cambridge, Mass.: Harvard University Press, 1979), pp. 6–7.

18. I am grateful to Professor Maynard Smith for pointing out to me that "short-term" and "long-term" memory are presumably different processes. One can lose the latter without losing the former, and the opposite process happens in some old people.

19. There is already considerable interest in the subject, for example G. Lynch and M. Baudry, "The Biochemistry of Memory: A New and Specific Hypothesis," *Science*, 224 (8 June 1984), 1057–1063; reprinted in P. H. Abelson, E. Butz, S. H. Snyder, eds., *Neuroscience* (Washington, D.C.: American Association for the Advancement of Science, 1985). "Our conclusion is that the mammalian brain does possess a chemical mechanism that could account for memory and yet is not likely to be involved in the ongoing operation of neuronal circuitries. The calcium proteinase-receptor process matches the conditions imposed by the behavioral features of memory." *Science*, p. 1062; *Neuroscience*, p. 415. See also J. Z. Young, *Philosophy and the Brain* (Oxford: Oxford University Press, 1987).

20. Jorge Luis Borges, "Funes, the Memorious," in *A Personal Anthology* (London: Jonathan Cape, 1968), p. 40.

21. A father has written about his own two young children: "But then all experience is exceptional to them; everything is new and distinct. The past closes up behind them immediately, while the future is an unimagined blank. They live on an edge of time. Like animals, they are absorbed in the present, in time being, and make no history of their own. Adult notions of the sequence of time and the cyclical ordering of existence have no meaning for them: all meals are 'supper' to Jack." F. Harrison, *A Father's Diary* (London: Fontana, 1985), p. 25.

22. Sigmund Freud, *The Unconscious*, The Complete Psychological Works of Sigmund Freud, Standard Edition, vol. 14 (London: Hogarth Press, 1957), p. 187.

23. Sigmund Freud, "The Dissection of the Physical Personality," New Introductory Lectures on Psychoanalysis, in *The Complete Introductory Lectures on Psycho-Analysis*, trans. and ed. James Strachey (Oxford: Alden and Mowbray, 1971), p. 538.

24. Dewey, *Human Nature and Conduct*, p. 70.

25. Ibid., p. 180.

26. Shakespeare, *Henry V*, 5.1.65.

27. David Hume, *Enquiries concerning Human Understanding and concerning the Principles of Morals* (Oxford: Clarendon Press, 1985), p. 43.

28. Ibid.

29. F. C. Bartlett, *Remembering: A Study in Experimental and Social Psychology* (Cambridge: Cambridge University Press, 1932), p. 309.

30. E. H. Carr, *What is History?* (London: Macmillan, 1961), p. 19.

31. A similar image was used before by myself and Willmott but for people drawn up in order of their standard of living, obeying the rule advanced by Daniel Bell that "what the few have today, the many will demand tomorrow." Daniel Bell, "The Year 2000: The Trajectory of an Idea," *Daedalus* (Summer 1967), 643. Quoted in M. Young and P. Willmott, *The Symmetrical Family* (London: Routledge and Kegan Paul, 1973; Penguin, 1984), p. 19.

32. Marcel Proust, *Remembrance of Things Past*, vol. 12, *Time Regained* (London: Chatto and Windus, 1941), p. 414.

33. A language changes continuously even though it does not go so far as to split off into a new one. For example, English established itself in something like its modern form by a large-scale change between the time of Chaucer and that of Shakespeare, called the Great Vowel Shift. "Before this time the vowel letters were pronounced as in German or Italian now . . . The first vowel letter in words like 'sane, sleep, five' is now pronounced with the post-Shift values, differently from 'sanity, slept and fifth.'" L. L. Cavalli-Sforza and M. W. Feldman, *Cultural Transmission and Evolution: A Quantitative Approach* (Princeton, N.J.: Princeton University Press, 1981), p. 21.

34. Arthur Koestler said of the biological parallel, "If evolution could only create novelties by starting each time afresh from the 'primeval soup,' the four thousand million years of the earth's history would not have been long enough to produce even an amoeba." "The Holon," in *Bricks to Babel* (London: Hutchinson, 1980), p. 460.

35. R. M. Titmuss, "The Position of Women," in his *Essays on the Welfare State* (London: Allen and Unwin, 1958), p. 91.

36. Larry Hirschorn, "Social Policy and the Life Cycle: A Developmental Perspective," *Social Service Review* (September 1977), 434–450. Quoted in B. L. Neugarten and D. A. Neugarten, "Age in the Aging Society," *Daedalus* (Winter 1986), 34.

37. By another kind of analogy, societies are frequently referred to as though they were people in their infancy, youth, maturity, old age. "'For America in her infancy to adopt the maxims of the Old World,' warned Noah Webster in his popular spelling book, 'would be to stamp the wrinkles of old age upon the bloom of youth and to plant the seed of decay in a vigorous constitution.'" W. A. Achenbaum, "America as an Aging Society: Myths and Images," *Daedalus* (Winter 1986), 17.

38. A. R. Radcliffe-Brown, *Structure and Function in Primitive Society* (London: Cohen and West, 1952), p. 10.

39. B. Quain, *Fijian Village* (Chicago, 1948). Quoted by C. Scott and G. Sabagh in "The Historical Calendar as a Method of Estimating Age," *Population Studies*, 24 (March 1970), 93–109.

40. Aristotle, *The Nicomachean Ethic*, ed. E. Capps, T. E. Page, and W. H. D. Rouse (London: Heinemann; New York: Putnam, 1926), book 2, p. 73: "Law givers make the citizens good by training them in habits of right

action—this is the aim of all legislation, and if it fails to do this it is a failure; this is what distinguishes a good form of constitution from a bad one."

41. Shakespeare, *Romeo and Juliet*, 1.3.23.

42. This usage is similar to that of Cavalli-Sforza and Feldman, *Cultural Transmission and Evolution*, pp. 54–55. According to these two biologists, transmission from parents to children is best considered *vertical*; from other people of an older generation (e.g., teachers) it is *oblique*; and from people of the offspring's own generation it is *horizontal*. They suggest that the importance of horizontal transmission is relatively new: "Cultural transmission must have been primarily vertical for much of human evolution. For more than 99% of human evolution social groups were of small size on average."

43. *Children and Their Primary Schools: A Report of the Central Advisory Council for Education, Department of Education and Science* (the Plowden Report) (London: Her Majesty's Stationery Office, 1967), pp. 136–137.

44. P. Willmott, *Adolescent Boys of East London* (London: Routledge and Kegan Paul, 1966), p. 31.

45. Karl Mannheim, "The Sociological Problem of Generations," in *Essays on the Sociology of Knowledge* (London: Routledge and Kegan Paul, 1952), p. 302.

46. James, *Principles of Psychology*, I, p. 121.

47. H. L. Wilensky, "Work, Careers, and Social Integration," *International Social Science Journal*, 12, no. 4 (1960), 554.

48. Young and Willmott, *The Symmetrical Family*, pp. 155–156.

49. Maurice Halbwachs, *The Collective Memory* (New York: Harper and Row, 1980), pp. 89–90. He also said: "Indeed, we become so used to measuring time in order to use it fully that we no longer know what to do with those portions of duration not so measured, when we are on our own and outside the current of external social life, as it were."

50. P. G. Zimbardo, C. Haney, W. Curtis Banks, and D. Jaffe, "The Psychology of Imprisonment: Privation, Power, and Pathology," in *Contemporary Issues in Social Psychology*, 4th ed., eds. J. C. Brigham and L. S. Wrightsman (Monterey, Calif.: Brooks-Cole, 1980), pp. 230–245. In another even more objectionable experiment at Yale volunteers were told to press a shock generator to give more and more severe electric shocks to victims whenever the victims made mistakes. The victims, who were in sight of the volunteers, pretended they were hurt and yelled and moaned. Although the volunteers believed they were causing the pain, they continued to obey the experimenter. Stanley Milgram, "Some Conditions of Obedience and Disobedience to Authority," *Human Relations*, 18 (1965), 57–75.

51. "The scope of determination of social relationships and cultural phenomena by authority and imperative co-ordination is considerably broader than appears at first sight. For instance, the authority exercised in the school has much to do with the determination of the forms of speech and of written language which are regarded as orthodox . . . The authority of parents and of the school, however, extends far beyond the determination of such cultural patterns which are perhaps only apparently formal, to the formation of the

character of the young, and hence of human beings generally." Weber, *The Theory of Social and Economic Organisation*, p. 327.

52. James, *Principles of Psychology*, I, p. 121.

53. It can be objected that the approach to habit is unduly holistic. Habit as a term does have a portmanteau quality, being almost synonymous in different contexts with personality and society. Yet I am sure that study of the parts is as necessary as study of the whole and so can echo for sociology what Maynard Smith has said about biology: "I have no doubt that the outstanding problems in biology will require such a dual attack: in contemporary jargon, both top down and bottom up. But to be neutral in such controversies is to invite the hatred and contempt of both sides." J. Maynard Smith, *The Problems of Biology* (Oxford: Oxford University Press, 1986), p. vi.

54. R. W. Emerson, "Circles," in *Selected Essays: Ralph Waldo Emerson*, ed. Larzer Ziff (New York: Penguin Books, 1982), p. 227.

55. There may here be another analogy with the biological process whereby repressor molecules can block the synthesis of messenger RNA from certain genes. Maynard Smith, *The Problems of Biology*, p. 66.

56. R. Dawkins, *The Selfish Gene* (New York: Oxford University Press, 1976), p. 206. Lumsden and Wilson use another word, "culturgens," and list others: the "mnemotype" of Blum, the "idea" of Huxley and Cavalli-Sforza, the "idene" of H. A. Murray, "sociogene" of Swanson, "instruction" of Cloak, "culture type" of Boyd and Richerson, and "concept" of Hill. Charles J. Lumsden and Edward O. Wilson, *Genes, Mind, and Culture: The Coevolutionary Process* (Cambridge, Mass.: Harvard University Press, 1981), p. 7.

57. Noam Chomsky, *Reflections on Language* (New York: Pantheon, 1976), p. 34.

58. Bruner suggests that a grammar also is developed in correspondence to the conceptual framework that is constructed by the repeated regulation of joint action and joint attention. Jerome S. Bruner, "The Ontogenesis of Speech Acts," *Journal of Child Language*, 2 (1975), 17.

59. P. Bateson, "Genes, Environment, and the Development of Behaviour," in T. R. Halliday and P. J. B. Slater, eds., *Animal Behaviour*, vol. 3, *Genes, Development and Learning* (Oxford: Blackwell, 1983) pp. 52–81.

60. Daniel J. Boorstin, *The Discoverers* (London: J. M. Dent, 1984), p. 388.

5. Social Evolution

1. Hannah Arendt, *The Human Condition* (Chicago: University of Chicago Press, 1958), p. 96.

2. Julian Huxley, ed., *The Humanist Frame* (New York: Harper Brothers; London: George Allen & Unwin, 1961), p. 13. In his book *Evolution: Modern Syntheses* (London: George Allen & Unwin, 1974), Huxley also says "After the disillusionment of the early twentieth century it has become as fashionable to deny the existence of progress and to brand the idea of it as a human illusion, as it was fashionable in the optimism of the nineteenth century to proclaim not only its existence but its inevitability. The truth is between the two extremes" (p. 578).

3. Teilhard de Chardin, *The Phenomenon of Man* (London: Collins, 1959), pp. 180–184.

4. Julian Huxley, *Man in the Modern World* (London: Chatto and Windus, 1947), p. 6.

5. The errors were not necessarily that many. "In genetic terms, we are hardly more distinct from chimpanzees than are subspecies in other groups of animals." Colin Patterson, *Evolution* (London: Routledge and Kegan Paul, 1978), p. 141.

6. Alfred Schutz and Thomas Luckmann, *The Structures of the Life-World* (Chicago: Northwestern University Press, 1973), p. 10.

7. Referred to in G. L. Stebbins, *Processes of Organic Evolution* (Englewood Cliffs, N.J.: Prentice-Hall, 1966), p. 175.

8. Francis Crick, *Life Itself: Its Origin and Nature* (London: Macdonald, 1982), p. 66.

9. E. B. Tylor, *Primitive Culture: Researches into the Development of Mythology, Philosophy, Religion, Art and Custom* (1871), quoted in Milton Singer, "The Concept of Culture," in *International Encyclopaedia of the Social Sciences*, ed. D. Sills (London: Macmillan, 1968), III, 527.

10. Alfred L. Kroeber and Talcott Parsons, "The Concept of Culture and of Social System," *American Sociological Review*, 23 (October 1958), 582.

11. Talcott Parsons, *The Structure of Social Action* (Glencoe, Ill.: The Free Press, 1949), p. 75.

12. According to E. H. Carr, "The cult of individualism is one of the most pervasive of modern historical myths. According to the familiar account in Burckhardt's *Civilization of the Renaissance in Italy*, the second part of which is subtitled 'The Development of the Individual,' the cult of the individual began with the Renaissance when man, who had hitherto been 'conscious of himself only as a member of a race, people, party, family, or corporation,' at length 'became a spiritual individual and recognised himself as such.' Later the cult was connected with the rise of capitalism and of Protestantism, with the beginnings of the industrial revolution, and with the doctrines of laissez-faire . . . But the whole process was a social process representing a specific stage in historical development, and cannot be explained in terms of a revolt of individuals against society or of an emancipation of individuals from social restraints." E. H. Carr, *What is History?* (London: Macmillan, 1961), pp. 27–28.

13. Alfred North Whitehead, *Science and the Modern World* (London: Cambridge University Press, 1926), p. 136.

14. "Whoever shall refuse to obey the general will must be constrained by the whole body of his fellow citizens to do so: which is no more than to say that it may be necessary to compel a man to be free—freedom being that condition which, by giving each citizen to his country, guarantees him from all personal dependence and is the foundation upon which the whole political machine rests, and supplies the power which works it." Jean-Jacques Rousseau, "The Social Contract," in *Essays by Locke, Hume, and Rousseau* (London: Oxford University Press, 1947), p. 261.

15. For this note I am indebted to my son, Toby Young.

16. "The structures of power may change quickly: new men arrive, new

routes of social ascent are opened, new bases of command created. Yet such dramatic overturns are largely a circulation of elites. Societal structures change much more slowly, especially habits, customs, and established, traditional ways . . . The time-dimensions of social change are much slower, and the processes more complex, than the dramaturgic mode of the apocalyptic vision, religious or revolutionary, would have us believe." Daniel Bell, *The Cultural Contradictions of Capitalism* (London: Heinemann, 1976), pp. 7–8. There are, however, examples that tell the other way. When faced by Western technology and values, tribal religions and rituals can wilt in a surprisingly short time. See V. W. Turner, *The Drums of Affliction* (Oxford: Clarendon, 1968), pp. 22–23.

17. Quoted in Daniel J. Boorstin, *The Discoverers* (London: Dent, 1984), p. 483.

18. Edward H. Levi, *An Introduction to Legal Reasoning* (Chicago: University of Chicago Press, 1949), p. 1.

19. Oliver Wendell Holmes, Jr., *The Common Law* (1881; Cambridge, Mass.: Harvard University Press, 1963), pp. 5, 8.

20. Richard Kluger, *Simple Justice: The History of Brown v. Board of Education and Black America's Struggle for Equality* (New York: Knopf, 1976). The changes that followed it were, of course, slow to arrive. "The decision was very controversial, the process of integration that followed was slow, and significant progress required many more legal, political and even physical battles." Ronald Dworkin, *Law's Empire* (London: Fontana, 1986), p. 30.

21. I. B. Cohen, *Revolution in Science* (Cambridge, Mass., and London: Harvard University Press, 1985), p. 18.

22. A cosmological physicist said to me that the only habit of mind he recognized was the habit of innovation. As long as one does not flatly contradict what has become known recently, the only rule in developing ideas is to be as outlandish as possible. A distinguished physicist, Ralph Leighton, said of another, "Richard Feynman is legendary in the world of physics for the way he looks at the world: taking nothing for granted and always thinking things out for himself, he often attains a new and profound understanding of nature's behaviour." In his preface to Richard P. Feynman, *QED: The Strange Theory of Light and Matter* (Princeton, N.J.: Princeton University Press, 1985).

23. J. M. Ziman, *Public Knowledge: An Essay concerning the Social Dimension of Science* (Cambridge: Cambridge University Press, 1968), p. 58.

24. Thomas S. Kuhn, *The Structure of Scientific Revolutions*, 2nd ed. (Chicago: University of Chicago Press, 1970), p. 23.

25. Ibid., p. 37.

26. Ibid., p. 96.

27. Ibid., p. 65.

28. Ibid., p. 53.

29. Richard Dawkins reports a similar reaction from a distinguished scientist, who could not believe Donald Griffin and Robert Galambos's report in 1940 of the discovery of the astonishing system of sonar "echolocation" used by bats. The scientist "seized Galambos by the shoulders and shook him while complaining that we could not possibly mean such an outrageous suggestion.

Radar and sonar were still highly classified developments in military technology, and the notion that bats might do anything even remotely analogous to the latest triumphs of electronic engineering struck most people as not only implausible but emotionally repugnant." Richard Dawkins, *The Blind Watchmaker* (London: Longman, 1986), p. 35.

30. Jerome S. Bruner and L. Postman, "On the Perception of Incongruity: A Paradigm," *Journal of Personality*, 18 (1949), 206–223.

31. David Cannadine, "The Context, Performance, and Meaning of Ritual: The British Monarchy and the 'Invention of Tradition' c. 1820–1977," in *The Invention of Tradition*, ed. Eric Hobsbawm and Terence Ranger (Cambridge: Cambridge University Press, 1983), pp. 101–164. The same book has a marvelous essay by Hugh Trevor-Roper, "The Invention of Tradition: The Highland Tradition of Scotland," about the retrospective invention of a distinct Highland culture and tradition, including the Highland dress and tartans.

32. Cannadine, "Context, Performance, and Meaning," p. 121.

33. Ernest Gellner, *Thought and Change* (London: Weidenfeld and Nicolson, 1964), p. 43.

34. "My view is that each particular mode of production, and the relations of production corresponding to it at each given moment, in short 'the economic structure of society,' is 'the real foundation on which arises a legal and political superstructure and to which correspond definite forms of social consciousness,' and 'the mode of production of material life conditions the general process of social, political and intellectual life,'" Karl Marx, *Capital*, vol. 1 (Harmondsworth: Penguin, 1976), p. 175. Marx is quoting his own preface to *A Contribution to the Critique of Political Economy*.

35. See J. L. Cloudsley-Thompson, "Environment and Human Evolution," *Environmental Conservation*, 2 (Winter 1975), 265. John Maynard Smith has pointed out that upright posture is also responsible for one of our "design faults": "For example, the bend in the lower spine of humans which is responsible for so many back pains exists because our ancestors were quadrupeds and we have only recently started standing upright." John Maynard Smith, *The Problems of Biology* (Oxford: Oxford University Press, 1986), p. 51. In evolutionary terms back pain could be considered a growing pain.

36. The terms are attributed to A. J. Lotka, by P. B. Medawar in *The Uniqueness of the Individual*, 2nd rev. ed. (New York: Dover, 1981), p. 119. See also Freud: "By means of all his tools, man makes his own organs more perfect—both the motor and the sensory—or else removes the obstacles in the way of their activity. Machinery places gigantic power at his disposal which, like his muscles, he can employ in any direction; ships and aircraft have the effect that neither air nor water can prevent his traversing them. With spectacles he corrects the defects of the lens in his own eyes; with the microscope he overcomes the limitations in visibility due to the structure of his retina." Sigmund Freud, *Civilization and Its Discontents* (London: Hogarth Press, 1930), p. 51.

37. Lenses have evolved by steps, like everything else. "Early lenses were all convex and so could assist only long-sighted people: the concave lenses, necessary for the short-sighted, came nearly two centuries later." T. K. Derry

and T. I. Williams, *A Short History of Technology from the Earliest Times to A.D. 1900* (Oxford: Clarendon, 1960), p. 112.

38. Of the many fluctuations up and down in economic activity, some of the most important have been the relatively long cycles triggered by new technologies. The greatest student of the Kondratieff cycle of 50 years or more, J. A. Schumpeter, has said, "Historically, the first Kondratieff covered by our material means the industrial revolution, including the protracted process of its absorption. We date it from the eighties of the eighteenth century to 1842. The second stretches over what has been called the age of steam and steel. It runs its course between 1842 and 1897. And the third, the Kondratieff of electricity, chemistry and motors, we date from 1898 on." J. A. Schumpeter, *Business Cycles* (New York: McGraw-Hill, 1939), p. 170.

39. H. S. Harrison, "Discovery, Invention, and Diffusion," in C. Singer, E. J. Holmyard, and A. R. Hall, eds., *A History of Technology,* vol. 1, *From Early Times to Fall of Ancient Empires* (Oxford: Clarendon, 1954), p. 62.

40. Patterson, *Evolution,* p. 56.

41. V. Gordon Childe, "Early Forms of Society," in *A History of Technology* I, 39. Braudel says that in Western Europe (and China) demographic cycles were succeeded by linear growth in about 1750. Since then, population "registers a continuous rise, more or less rapid according to society and economy but always continuous. Previously it rose and then fell like a series of tides." There had been a prolonged population rise between 1100 and 1350, then a regression; then another rise between 1450 and 1650 also followed by a regression. These fluctuations were perhaps due to climatic change, to which people were far from immune before the large increases in food supply brought by the industrial revolution. Fernand Braudel, *Civilization and Capitalism,* vol. 1, *The Structures of Everyday Life* (London: Collins, 1981), pp. 31–33, 49.

42. Aldous Huxley, "Human Potentialities," in *The Humanist Frame,* ed. Julian Huxley (New York: Harper, 1961), p. 423.

43. Medawar, *Uniqueness of the Individual,* pp. 121–122.

44. Darwin, in a section in *Origin of Species* on the degree to which organization tends to advance, put the point this way: "If we take as the standard of high organisation, the amount of differentiation and specialisation of the several organs in each being when adult (and this will include the advancement of the brain for intellectual purposes), natural selection clearly leads towards this standard: for all physiologists admit that the specialisation of organs, inasmuch as in this state they perform their functions better, is an advantage to each being; and hence the accumulation of variations tending towards specialisation is within the scope of natural selection." Charles Darwin, *Origin of Species* (New York: New American Library, 1958), p. 123.

45. E. P. Thompson, *The Making of the English Working Class* (London: Gollancz, 1963), p. 416.

46. Dawkins, *The Blind Watchmaker,* p. 90.

47. Edward Gibbon, *The Decline and Fall of the Roman Empire* (London: Dent, 1910), IV, 112. Quoted in Carr, *What is History?,* p. 111.

48. J. B. Bury, *The Idea of Progress* (London: Macmillan, 1920), pp. vii–viii.

49. M. J. Caritat, Marquis de Condorcet, *Outlines of an Historical View of the Progress of the Human Mind* (London, 1795). Quoted in Sidney Pollard, *The Idea of Progress* (London: Watts, 1968), p. xii.

50. "Part of the fortunate conjunction of circumstance with respect to us who live here in the United States consists, as has been indicated, of the fact that our forefathers found themselves in a new land. The shock of physical dislocation effected a very considerable modification of old attitudes. Habits of thought and feeling which were the products of long centuries of acculturation were loosened. Less entrenched dispositions dropped off. The task of forming new institutions was thereby rendered immensely easier . . . It is because of such consequences that the geographical New World may become a New World in a human sense." John Dewey, *Freedom and Culture* (New York: Putnam, 1939), p. 174.

51. Gellner, *Thought and Change,* p. 39.

52. The British Empire was still imposed with a great deal of ruthlessness. "Where the transition is rapid, the effect is acute and is felt by great masses of people. World history offers no spectacle more frightful than the gradual extinction of the English hand-loom weavers; this tragedy dragged on for decades, finally coming to an end in 1838. Many of the weavers died of starvation, many vegetated their families for a long period on 2-1/2 d a day. In India, on the other hand, the English cotton machinery produced an *acute* effect. The Governor General reported as follows in 1834–5. 'The misery hardly finds a parallel in the history of commerce. The bones of the cotton-weavers are bleaching the plains of India.'" Karl Marx, *Capital* (Harmondsworth: Penguin, 1976), I, 558.

53. The same phenomenon is evident even in the countries to which power has passed. "To this domestic agenda has been added a host of issues resulting from the active international role the United States has pursued since World War II. As our interdependence with other parts of the world has grown, fewer and fewer important activities can be regarded any longer as purely domestic in nature . . . Problems of pollution, disease, population, drugs, and many more overreach national boundaries . . . Beyond these immediate problems lies the ultimate issue of nuclear weapons, which links us to all peoples throughout the world." Derek Bok, *Higher Learning* (Cambridge, Mass.: Harvard University Press, 1986), pp. 164–165.

54. J. W. Burton, *World Society* (Cambridge: Cambridge University Press, 1972), pp. 32–33.

55. Much has clearly happened to the language since 1582, when Richard Mulcaster said that "the English tongue was of small reach stretching no farther than this island of ours and not there over all." Quoted in Harold Goad, *Language in History* (Harmondsworth: Penguin, 1958), p. 185.

56. *The Right Word at the Right Time* (London: The Readers Digest Association, 1985), pp. 214–216.

57. Some from Africa are *like a bushfire in the harmattan, like a yam tendril*

in the rainy season, where there is dew there is water, wisdom is like a goat skin—everyone carries his own, like a lizard fallen from an iroke tree, like pouring grains of corn into a bag full of holes, to eat other's ears (to talk privately), *to have no shadow* or *to have no bite* (to have no courage). See B. B. Kachru, "Standards, Codification, and Sociolinguistic Realisms: The English Language in the Outer Circle," in *English in the World: Teaching and Learning the Language and Literatures,* ed. Randolph Quirk and H. G. Widdowson (Cambridge: Cambridge University Press, 1985), p. 19.

58. *The Economist* quoted an announcement on a cross-Channel ferry: "Ladies and gentlemen, mesdames et messieurs: the buffet is now open, le snackbar est maintenant ouvert." "The New English Empire," *The Economist* (20 Dec.–2 Jan. 1987), p. 129.

59. *The Right Word at the Right Time,* p. 215.

60. It is, perhaps, just conceivable that the outcome will be as it was before: "And the whole earth was of one language, and they used few words. And they found a plain in the land of Shinar; and they settled there. And they said to one another, Go to, let us make bricks, and burn them hard. And they had brick for stone, and slime they had for mortar. And they said, Go to, let us build us a city and a tower whose top may reach unto heaven and make a name for ourselves. And the Lord came down to see the city and the tower, which the children of men builded; and the Lord said, Behold, they are one people, and they have all one language, and now they have started to do this; henceforward nothing they have a mind to do will be beyond their reach." Arthur Koestler, *Bricks to Babel* (London: Hutchinson, 1980), pp. 684–685.

61. Quirk and Widdowson, *English in the World,* p. 3.

62. Alfred North Whitehead, quoted in Medawar, *Uniqueness of the Individual,* p. 117.

63. I am grateful to Simon Szreter for his help with the observation of the newsroom.

64. Asa Briggs, *The History of Broadcasting in the United Kingdom,* vol. 2, *The Golden Age of Wireless* (London: Oxford University Press, 1965), p. 154.

65. Claude Lévi-Strauss, *The View from Afar,* trans. J. Neugroschel and P. Hass (Oxford: Blackwell, 1985), p. 23.

66. V. I. Lenin, *The State and Revolution* (Moscow: Progress Publishers, 1965), p. 96.

67. Émile Durkheim, *The Division of Labor in Society* (New York: Free Press, 1964).

68. See Dawkins, *The Blind Watchmaker.*

69. M. F. Ashley Montagu, *The Direction of Human Development: Biological and Social Bases* (London: Watts, 1957), p. 27. These principles have not yet been converted into industrial and commercial practice on any scale. See Michael Young and Marianne Rigge, *Revolution from Within* (London: Weidenfeld and Nicolson, 1983).

70. T. H. Marshall, *Citizenship and Social Class* (Cambridge: Cambridge University Press, 1950).

71. Michel Foucault, *Discipline and Punish* (London: Allen Lane, 1977), p. 3.

72. Nadezhda Mandelstam, *Hope Abandoned: A Memoir,* trans. M. Hayward (Harmondsworth: Penguin Books, 1976).

73. J. T. Fraser, ed., *Voices of Time,* 2nd ed. (Amherst: University of Massachusetts Press, 1981), p. xxx.

6. The Metronomic Pulse

1. G. J. Whitrow, *The Natural Philosophy of Time* (Oxford: Clarendon Press, 1980), chap. 2.

2. P. A. Sorokin and R. K. Merton, "Social Time: A Methodological and Functional Analysis," *American Journal of Sociology,* 42 (March 1937), 623.

3. S. G. F. Brandon, "Time and the Destiny of Man," in *The Voices of Time,* ed. J. T. Fraser (London: Allen Lane, 1968), p. 149.

4. David D. Landes, *Revolution in Time* (Cambridge, Mass., and London: Harvard University Press, 1983), p. 150.

5. Daniel J. Boorstin, *The Discoverers* (London: Dent, 1984), p. 39. Until the fourteenth century the kind of hours had to be specified as well as how many of them. In the reign of Valentinian I (364–375), Roman soldiers were drilled to march "at the rate of 20 miles in five *summer* hours" (p. 28).

6. It has been suggested that there was originally no perspective in painting because space and time were not differentiated. Jesus could appear three times over in the same picture. Perspective was an acceptance that things coexisted— that you could be aware of many different things happening simultaneously and order that awareness. A. Appels, "Cultural Aspects of Coronary Prone Behaviour: An Approach from the History of Art," in *Advances in Cardiology,* vol. 29, *Psychological Problems before and after Myocardial Infarction* (Basel: S. Karger, 1982), pp. 32–36.

7. Landes, *Revolution in Time,* p. 95.

8. The anecdote is from Gislebert of Mons. Marc Bloch, *Feudal Society* (London: Routledge and Kegan Paul, 1961), I, 73–74, quoted in W. G. Runciman, *A Treatise on Social Theory* (Cambridge: Cambridge University Press, 1983), I, 261.

9. Max Weber, "Protestant Asceticism and the Spirit of Capitalism," in *Max Weber: Selections in Translation,* ed. W. G. Runciman (Cambridge: Cambridge University Press, 1978), p. 141.

10. Hardy's Dorset people would have been sympathetic. "The next evening the mummers were assembled in the same spot, awaiting the entrance of the Turkish Knight. 'Twenty minutes after eight by the Quiet Woman, and Charley not come.' 'Ten minutes past by Blooms-End.' 'It wants ten minutes to, by Grandfer Cantle's watch.' 'And 'tis five minutes past by the captain's clock.' . . . West Egdon believed in Blooms-End time, East Egdon in the time of the Quiet Woman Inn, Grandfer Cantle's watch had numbered many followers in years gone by, but since he had grown older faiths were shaken. Thus, the mummers

having gathered hither from scattered points, each came with his own tenets on early and late; and they waited a little longer as a compromise." Thomas Hardy, *The Return of the Native* (London: Macmillan, 1974), p. 154.

11. Pierre Bourdieu, "The Attitude of the Algerian Peasant toward Time," trans. G. E. Williams, in *Mediterranean Countryman*, ed. Julian Pitt-Rivers (Paris: Mouton, 1963), p. 57.

12. E. T. Hall, *The Silent Language* (New York: Doubleday, 1959), p. 26.

13. "It is perhaps important to consider that a nun who spends sixty years in a Carmelite enclosure, if she is in the chapel choir for Mass and Readings at 8 am and 8 pm every day, is in the same place at the same time exactly 42,800 times during those years." Drid Williams, "The Brides of Christ," in *Perceiving Women*, ed. S. Ardener (London: Dent; New York: Wiley, 1975), p. 115. Similarly, a Benedictine monk who lasted as long could have heard the same bells calling him to prayer not far from two hundred thousand times.

People can impose the same discipline upon themselves outside monasteries, as W. H. Auden "By conforming to a rigid time schedule, Auden was exerting his control over time as surely as did those ancient Benedictine monks in the Cluniac and Cistercian monasteries of France—Saint-Guilhem-le-Desert, Paray-le-Monial, Anay-le-Duc, Fontevrault, Vezelay—who, living by the canonical hours, were asserting their will over time in order to live in eternity . . . Most of the time Auden lived a monastic time schedule from lauds to compline. Usually he arose before six and preferred to retire by nine or ten. He worked at his writing from prime through sext." Dorothy J. Farnan, *Auden in Love* (London: Faber and Faber, 1985), pp. 175–176.

14. Landes, *Revolution in Time*, pp. 60–66.

15. Lewis Mumford, *Technics and Civilization* (New York: Harcourt Brace, 1934), pp. 13–17.

16. Jean Piaget, *The Child's Conception of Time* (London: Routledge and Kegan Paul, 1969), p. 250.

17. Tun Li-ch'en, *Annual Customs and Festivals in Peking*, trans. Derek Bodde (Hong Kong: Hong Kong University Press, 1965), p. 106.

18. "In some tribes . . . the 'depth' of genealogies (i.e. the number of supposed ancestors) remains the same, however many generations pass: identities and relationships of the dead and of the living in a sense 'change' (by our criteria); some ancestors are forgotten, and the present generation has the 'same' relationship to the permanently remembered founders, etc., as had the preceding generation; so that the pattern of an eternal present and of its temporal horizon remain ever the same." Ernest Gellner, *Thought and Change* (London: Weidenfeld and Nicolson, 1964), p. 2. Evans-Pritchard said of the Nuer in Africa, "Valid history ends a century ago, and tradition, generously measured, takes us back only 10 to 12 generations in lineage structure, and if we are right in supposing that lineage structure never grows, it follows that the distance between the beginning of the world and the present remains unalterable." Edward Evans-Pritchard, *The Nuer* (Oxford: Clarendon, 1940), p. 108.

19. Elizabeth Eisenstein, *The Printing Revolution in Early Modern Europe* (Cambridge: Cambridge University Press, 1983), p. 272.

20. Ibid., p. 275.

21. Hans Meyerhoff, *Time in Literature* (Berkeley and Los Angeles: University of California Press, 1955), p. 109.

22. Maurice Halbwachs, *The Collective Memory* (New York: Harper and Row, 1980), p. 52.

23. H. F. Amiel, *Jean-Jacques Rousseau* (New York: B. W. Huebsch, 1922), quoted in R. D. Masters, "Jean-Jacques Is Alive and Well: Rousseau and Contemporary Sociobiology," *Daedalus* (Summer 1978), 93–105.

24. "It takes time for the absent to assume their true shape in our thoughts. After death they take on a firmer outline and then cease to change. 'So that's the real you? Now I see, I never understood you.' It is never too late, since now I have fathomed what formerly my youth hid from me: my brilliant, cheerful father harbored the profound sadness of those who have lost a limb." Colette, *Earthly Paradise* (London: Secker and Warburg, 1966), p. 53.

25. Eisenstein, *The Printing Revolution*, p. 35.

26. Ithiel de Sola Pool, Hiroshi Inose, Nozomu Takasaki, and Roger Hurwitz, *Communication Flows: A Census in the United States and Japan* (North Holland, Netherlands: University of Tokyo Press, 1984), pp. 10–11, 16–17.

27. Each town used to have its own time zone. When the first regular stagecoach services were started in England in 1784 they had to run through many different time zones, and coach passengers gained or lost time according to their direction just as jet passengers do now on a larger scale. Because local Bristol time differed from London by twenty minutes, the coach guard "would simply adjust his timepiece so that it lost twenty minutes on the 'down' run to London and gained twenty minutes on the 'up' run to Bristol." Nigel Thrift, "Owners' Time and Own Time: The Making of a Capitalist Time Consciousness, 1300–1880," in *Space and Time in Geography*, ed. Alan Pred (Lund, Sweden: Gleerup, 1981), p. 69.

28. Evans-Pritchard, *The Nuer*, p. 101.

29. R. A. Hahn, "A World of Internal Medicine: Portrait of an Internist," in *Physicians of Western Medicine*, ed. R. A. Hahn and A. D. Gaines (Dordrecht, Netherlands: Reidel, 1985), pp. 51–111. I am grateful to Ronald Frankenberg for drawing my attention to this account.

30. See G. J. Whitrow, *The Natural Philosophy of Time* (Oxford: Clarendon, 1980), p. 30, for this reference and for a discussion of what Bertrand Russell called the "Tristram Shandy paradox."

31. The preoccupation with time is shown by the titles of books in print in the United States in 1986: *Mastery and Management of Time; About Time: A Woman's Guide to Time Management; I Just Need More Time; Getting More Done in Less Time; How to Put More Time in Your Life; Tested Tactics That Conserve Time for Top Executives; Tips on Managing Your Time and Increasing Your Effectiveness.*

32. "In America I have seen the freest and best educated of men in circumstances the happiest to be found in the world; yet it seemed to me that a cloud habitually hung on their brow, and they seemed serious and almost sad even in their pleasures . . . [They] never stop thinking of the good things they have not got." Alexis de Tocqueville, *Democracy in America*, trans. G. Lawrence, ed. J. P. Mayer (New York: Doubleday, 1969), p. 536.

33. Staffan Linder, *The Harried Leisure Class* (New York: Columbia University Press, 1970), p. 79.

34. I am told that in California more people who suffer from burnout because they cannot stand the pressure are leaving the cities. See R. L. Meier, *A Communications Theory of Urban Growth* (Cambridge, Mass.: MIT Press, 1962).

35. "The United States is a relatively uncrowded country by world standards, but already our fastest growth is in regions (Arizona, New Mexico and other desert areas) which would not attract large numbers of people if there were no air conditioning." Gerard O'Neill, *The High Frontier: Home Colonies in Space* (London: Jonathan Cape, 1977), p. 40.

36. Karl Marx, *Capital* (Harmondsworth, England: Penguin Books, 1976), I, 367.

37. Murray Melbin, "Night as Frontier," *American Sociological Review,* 43 (Feb. 1978), 5.

38. Ibid., p. 12, quoting W. E. Hollon, *Frontier Violence* (New York: Oxford University Press, 1974), pp. 211–212.

39. These subjects are like more primitive people, who "sleep largely in the open, and are more familiar with the night sky than their civilized contemporaries. Nearly all have names for a few easily recognized stars or star clusters." E. R. Leach, "Primitive Time Reckoning," in *History of Technology,* vol. 1, *From Early Times to Fall of Ancient Empires,* ed. C. Singer, E. J. Holmyard, and A. R. Hall (Oxford: Clarendon, 1954), pp. 110–127.

40. See Michael Young and Peter Willmott, *The Symmetrical Family* (London: Routledge and Kegan Paul, 1973), pp. 251–254.

7. All Days Are Not Equal

1. Immanuel Kant, *Critique of Pure Reason* (London: Dent, 1956), p. 47.

2. Kathleen Raine, "The Moment," in *Collected Poems 1935–80* (London: George Allen and Unwin, 1981), p. 72.

3. Saint Augustine, *Confessions,* trans. R. S. Pine-Coffin (London: Penguin Books, 1961), p. 262.

4. See David Layzer, "The Arrow of Time," *Scientific American,* 233 (December 1975), pp. 56–69.

5. Steven Weinberg, *The First Three Minutes* (London: Flamingo, 1983), p. 20.

6. As one physicist has said, "Arrayed in an apparently random pattern in the sky, the stars provide a perfect screen for the projection of our feelings. In that pattern, ancient priests and poets saw the figures of myth and nature; the stars were gods—archetypes of permanence in an impermanent world. Compared with human life or the life of nations and empires, stars appear to live forever, indifferent to the passions of our existence. Yet somehow we feel that in spite of the immense distances which separate us from all stars save our sun, the destiny of humanity is profoundly intertwined with them. We hope that life on earth may share in the permanence of the stars, the galaxies and the universe itself. Whether that hoped-for permanence is no more than a projection upon

the heaven of our modern myth of progress and therefore, like the ancient projections of the figures of myth, also an illusion, time will tell. The stars, like the gods they once represented, continue to play with our deepest feelings." H. R. Pagels, *Perfect Symmetry: The Search for the Beginning of Time* (London: Michael Joseph, 1985), pp. 30–31.

7. This slow motion may not appear to be common. "Small mammals tick fast, burn rapidly, and live for a short time; large mammals live long at a stately pace. Measured by their own internal clocks, mammals of different sizes tend to live for the same amount of time." Stephen Jay Gould, "Our Allotted Lifetimes," in *The Panda's Thumb* (New York: Norton, 1982), p. 302.

8. Weinberg, *The First Three Minutes*, p. 49.

9. "Decision, as all of us use the word, is a cut between past and future, an introduction of an essentially new strand into the emerging pattern of history." G. L. S. Shackle, *Decision, Order, and Time in Human Affairs*, 2nd ed. (Cambridge: Cambridge University Press, 1969), p. 3.

10. Nigel Calder, *Timescale* (London: Hogarth, 1984), p. 18.

11. Richard Dawkins, *The Selfish Gene* (Oxford: Oxford University Press, 1976), p. 21.

12. John Ziman, *Public Knowledge: An Essay Concerning the Social Dimension of Science* (Cambridge: Cambridge University Press, 1968), p. 103.

13. Ira Progoff, *Jung's Psychology and Its Social Meaning* (London: Routledge and Kegan Paul, 1953), p. 232.

14. There is, perhaps, some comfort to be had from the fact that these fears were most eloquently expressed by the great Swiss historian, Jacob Burckhardt, over a century ago, in 1852. Jacob Burckhardt, *The Age of Constantine the Great* (London: Routledge and Kegan Paul, 1949).

15. Progoff, *Jung's Psychology*, p. 220.

16. Geoffrey Vickers, "Ecology, Planning, and the American Dream," in *The Urban Condition*, ed. L. J. Duhl (New York: Basic Books, 1963), p. 383.

17. Peter Berger and Thomas Luckmann, *The Social Construction of Reality* (Garden City, N.Y.: Anchor, 1967), p. 106.

18. Hilda Nowotny, "The Public and Private Uses of Time," in Laura Balbo and Hilda Nowotny, eds., *Time to Care in Tomorrow's Welfare Systems* (Vienna: European Centre for Social Welfare Training and Research, 1986), p. 14.

19. *The Anatomy of Melancholy,* quoted in T. A. Wehr and F. K. Goodwin, "Biological Rhythms and Psychiatry," in Silvano Arieti and H. K. H. Brodie, eds., *American Handbook of Psychiatry,* vol. 8 (New York: Basic Books, 1981), p. 46.

20. Émile Durkheim, *Suicide,* trans. J. Spaulding and G. Simpson (London: Routledge and Kegan Paul, 1952), pp. 106–109.

21. N. E. Rosenthal, D. A. Sack, and T. A. Wehr, "Seasonal Variation in Affective Disorders," in T. A. Wehr and F. K. Goodwin, eds., *Circadian Rhythms in Psychiatry* (Pacific Grove, Cal.: Boxwood Press, 1983), p. 187.

22. E. Corfman, *Depression, Manic-Depressive Illness, and Biological Rhythms,* Science Report 1 (Rockville, Md.: National Institute of Mental Health, 1979). T. A. Wehr and F. K. Goodwin say of diseases of the circadian system, "It seems likely that such diseases would affect multiple systems, be associated

with disruption of function rather than tissue, be of obscure etiology, and have prominent behavioral manifestations, including disturbances of the sleep-wake cycle. In a very general way, this description fits some psychiatric illnesses, especially the affective illnesses—depression and manic-depressive states." "Biological Rhythms and Psychiatry," in Wehr and Goodwin, *Circadian Rhythms in Psychiatry*, pp. 46–47.

23. Jacob A. Arlow, "Scientific Cosmogony, Mythology, and Immortality," *Psychoanalytic Quarterly*, 51 (1982), 177–195; and "Depersonalization and Derealization," in *Psychoanalysis: A General Psychology*, ed. H. Hartmann (New York: 1966). A provoking agent for depression can be a life event such as death, separation, loss of job, or loss of hope, which disturbs the pattern set up in the past, and threatens to continue to do so. See G. W. Brown and T. Harris, *Social Origins of Depression: A Study of Psychiatric Disorder in Women* (London: Tavistock, 1978).

24. F. T. Melges, *Time and the Inner Future: A Temporal Approach to Psychiatric Disorders* (New York: Wiley, 1982).

25. W. J. M. Hrushesky, "Circadian Timing of Cancer Chemotherapy," *Science*, 228 (5 April 1985), 73–75.

26. Having the same circadian propensities as other animals can be an advantage, too. It has led to a new therapy for children and old people, in and out of hospitals. "Four-footed therapists" can provide the elderly with companionship, something that keeps them active, something that makes them feel safe, something to touch, something to care for and, above all, something that gives regularity to life from one day to another. In one hospital study, in Ohio, the levels of medication, incidence of violence, and attempted suicide in a control ward without pets were double those in an experimental ward with pets present. D. R. Lee, "Pet-Therapy: Helping Patients through Troubled Times," *California Veterinarian*, 37 (1983), 24–25.

27. See Michael Young and Peter Willmott, *The Symmetrical Family* (London: Routledge and Kegan Paul, 1973), p. 175.

28. Simon Folkard and T. H. Monk, eds., *Hours of Work: Temporal Factors in Work Scheduling* (Chichester, England: Wiley, 1985).

29. C. A. Czeisler, M. C. Moore-Ede, and R. M. Coleman, "Rotating Shift Work Schedules That Disrupt Sleep and Are Improved by Applying Circadian Principles," *Science*, 217 (30 July 1982), 460–463.

30. I. H. Pearse and L. H. Crocker, *The Peckham Experiment* (London: Allen and Unwin, 1943), p. 82.

31. "From all that is known about the quality of experiences on the job it is on the lowest level of the industrial or bureaucratic hierarchy that traditional organisational forms are often experienced as oppressive. In contrast to the lack of time structure in the lives of the unemployed, time here is often too rigidly structured in minute periods in repetitive activities; there is no scope for initiative or self-determination; the purposes of activities and the manner in which they have to be carried out are often obscure; many regulations emphasise the low status of the employed that is heightened by a built-in mistrust of the ability of the employed to use judgement." Marie Jahoda, *Employment and Unemploy-*

ment: A Social Psychological Analysis (Cambridge: Cambridge University Press, 1982), pp. 68–69.

32. The term "Type A" originated in R. H. Rosenman et al., "Coronary Heart Disease in the Western Collaborative Group Study," *Journal of Chronic Diseases,* 23 (1970), 173–190.

33. R. H. Rosenman, "Current status of risk factors and type A behaviour pattern in the pathogenesis of ischemic heart disease," in T. M. Dembroski et al., *Biobehavioural Bases of Coronary Heart Disease* (Basel: Karger, 1983), pp. 5–17.

34. M. G. Marmot, M. J. Shipley, and G. Rose, "Inequalities in Death-Specific Explanations of a General Pattern?" *Lancet* (5 May 1984), 1003.

35. L. Alfredsson, R. Karasek, and T. Theorell, "Myocardial Infarction Risk and Psychosocial Work Environment: An Analysis of the Male Swedish Working Force," *Social Science and Medicine,* 16, no. 1 (1982) 463–467.

36. "While sudden changes and overstimulation commonly have a variety of traumatic effects, deprivation of stimuli seems paradoxically to be becoming a more common cause of pathology under modern conditions of work. The reason is that a certain amount of diversity is essential for mental well-being and probably also for physical health . . . Needless to say, the physiological and psychological effects of sensory deprivation have their counterpart in the disturbances associated with the monotony of automated work and dial-watching tasks." Rene Dubos, "Man Adapting to Working Conditions," in *Society, Stress, and Disease,* vol. 4, *Working Life,* ed. L. Levi (Oxford: Oxford University Press, 1981), pp. 4–5.

37. Richard Dawkins, *The Blind Watchmaker* (London: Longman, 1986), p. 94.

38. "Dr. Samuel Johnson had many lean and hungry years in London; and the story is told of an occasion when he was invited to visit his erstwhile pupil David Garrick. David Garrick was by then a most successful actor manager, and when Dr. Johnson saw the fine apartments, the lovely furniture, the books and the furs and the beautiful women, he said 'Davey, Davey, this must make death very terrible.' It takes the articulate honesty of a great man to react like that, yet two centuries later we need reminding. We have moved from a high morbidity, high fertility society to a low morbidity, low fertility one. The death of a baby is no longer accepted as the will of God but as a possible cause of litigation. In our suburbanised and centrally heated lives, death is more than ever terrible." E. Wilkes, "Overview," in *Proceedings of the Conference on Care of the Dying* (London: Her Majesty's Stationery Office, 1986), p. 9. This official British publication began with the disclaimer that the views expressed were not necessarily those of the Department of Health and Social Security who sponsored the Conference.

39. A reminder is in Samuel Pepys' seventeenth-century diary. His account is so engaging because he brought so much gusto to each new moment—every pretty woman seen through the window of his coach; each new dish of beef or even pease porridge; each song he sang, from a new composition of Purcell senior to a tavern ballad; and always the fact that every moment was different.

London was so small that he could walk around a different part of the city every day, or take a boat down the Thames, from Whitehall to the Navy Yard at Deptford or to the Tower. The diary is remarkable for what he had, and for what he did not have. He and his fellows did not have clocks as their constant companions. The time was tolled for them by the well-recognized bells of the churches, but not much attention had to be paid even to them. If Pepys arrived at the house of the Duke of York to find the Duke had gone hunting, he was not put out but sat down to talk and drink and let the time "pass" without any sense of "wasting" it. Pepys had a trying job in the administration of the Navy, but its discharge was mixed up with a hundred other interests and vocations which allowed him so often at the end of another day to go "away home to bed, with infinite content at the treat, for it was mighty pretty and everything mighty rich." See Arthur Bryant, *Samuel Pepys: The Man in the Making* (London: Panther, 1984).

40. Thomas Gradgrind in Charles Dickens, *Hard Times* (Basingstoke, England: Macmillan, 1983), pp. 79–80. It is the opposite of the response to the seasons of nature and the church described by D. H. Lawrence: "This is the wheeling of the year, the movement of the sun through solstice and equinox, the coming of the seasons, the going of the seasons. And it is the inward rhythm of man and woman, too, the sadness of Lent, the delight of Easter, the wonder of Pentecost, the fires of St. John, the candles on the grave of All Souls, the lit-up tree of Christmas, all representing kindled rhythmic emotions in the souls of men and women." D. H. Lawrence, "A Propos of Lady Chatterley's Lover," in *A Selection from Phoenix* (Harmondsworth, England: Penguin Books, 1971), p. 347.

41. "The work pattern was one of alternate bouts of intense labour and of idleness, whenever men were in control of their own working lives. On Monday or Tuesday, according to tradition, the hand-loom went to the slow chant of *Plen-ty of Time, Plen-ty of Time*: on Thursday and Friday, *A day t'lat, A day t'lat*." E. P. Thompson, "Time, Work-Discipline, and Industrial Capitalism," *Past and Present*, 38 (1967), 73.

42. See J. A. Arlow, "Disturbances of the Sense of Time," *Psychoanalytic Quarterly*, 53 (1984), 13–37.

Index